P9-APA-595

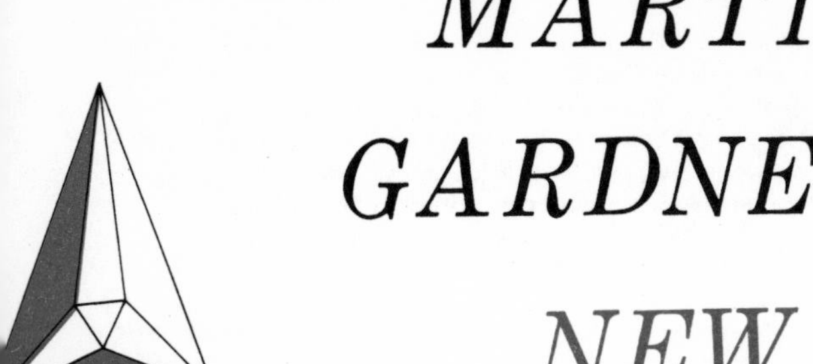

MARTIN GARDNER'S NEW MATHEMATICAL DIVERSIONS FROM SCIENTIFIC AMERICAN

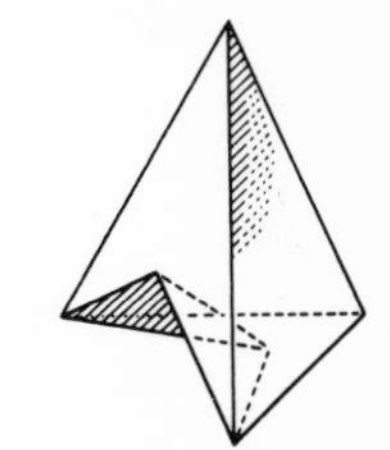

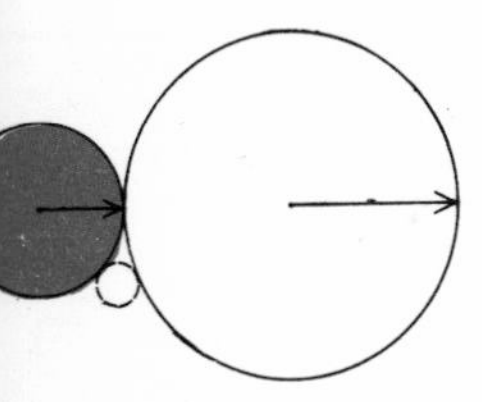

LONDON
GEORGE ALLEN AND UNWIN LTD

FIRST PRINTED IN GREAT BRITAIN IN 1969

SBN 04 510033 0

Most of the drawings and diagrams appear by courtesy of *Scientific American*, in whose pages they were originally published.

PRINTED IN GREAT BRITAIN
BY PHOTOLITHOGRAPHY BY
BUTLER & TANNER LTD
FROME, SOMERSET

Evoly met

TO L. R. AHCROF,

emitero meno.

CONTENTS

INTRODUCTION 9

1. The Binary System 13
[Answers on page 22]

2. Group Theory and Braids 23
[Answers on page 33]

3. Eight Problems 34
[Answers on page 39]

4. The Games and Puzzles of Lewis Carroll 47
[Answers on page 53]

5. Paper Cutting 58
[Answers on page 69]

6. Board Games 70
[Answers on page 82]

7. Packing Spheres 84
[Answers on page 92]

8. The Transcendental Number Pi 93
[Answers on page 102]

9. Victor Eigen: Mathemagician 105
[Answers on page 114]

10. The Four-Color Map Theorem 115
[Answers on page 124]

11. Mr. Apollinax Visits New York 126
[Answers on page 134]

12. Nine Problems 136
[Answers on page 142]

13. Polyominoes and Fault-Free Rectangles 152
[Answers on page 160]

14. Euler's Spoilers: The Discovery of an Order-10 Graeco-Latin Square 164
[Answers on page 173]

15. The Ellipse 175
[Answers on page 185]

16. The 24 Color Squares and the 30 Color Cubes 186
[Answers on page 197]

17. H.S.M. Coxeter 198
[Answers on page 211]

18. Bridg-it and Other Games 212
[Answers on page 218]

19. Nine More Problems 221
[Answers on page 227]

20. The Calculus of Finite Differences 236
[Answers on page 246]

REFERENCES FOR FURTHER READING 249

INTRODUCTION

"A good mathematical joke," wrote the British mathematician John Edensor Littlewood (in the introduction to his Mathematician's Miscellany), *"is better, and better mathematics, than a dozen mediocre papers."*

This is a book of mathematical jokes, if "joke" is taken in a sense broad enough to include any kind of mathematics that is mixed with a strong element of fun. Most mathematicians relish such play, though of course they keep it within reasonable bounds. There is a fascination about recreational mathematics that can, for some persons, become a kind of drug. Vladimir Nabokov's great chess novel, The Defense, *is about such a man. He permitted chess (one form of mathematical play) to dominate his mind so completely that he finally lost contact with the real world and ended his miserable life-game with what chess problemists call a suimate or self-mate. He jumped out of a window. It is consistent with the steady disintegration of Nabokov's chess master that as a boy he had been a poor student, even in mathematics, at the same time that he had been "extraordinarily engrossed in a collection of problems entitled* Merry Mathematics, *in the fantastical misbehavior of numbers and the wayward frolic of geometric lines, in everything that the schoolbook lacked."*

The moral is: Enjoy mathematical play, if you have the mind and taste for it, but don't enjoy it too much. Let it provide occasional holidays. Let it stimulate your interest in serious science and mathematics. But keep it under firm control.

And if you can't keep it under control, you can take some comfort from the point of Lord Dunsany's story "The Chess-Player, the Financier, and Another." A financier recalls a friend named Smoggs who was on the road to becoming a brilliant financier until he got sidetracked by chess. "It came gradually at first: he used to play chess with a man during the lunch hour, when he and I both worked for the same firm. And after a while he began to beat the fellow. . . . And then he joined a chess club, and some kind of fascination seemed to come over him; something like drink, or more like poetry or music . . . he could have been a financier. They say it's no harder than chess, though chess leads to nothing. I never saw such brains wasted."

"There are men like that," agrees the prison warden. "It's a pity . . ." And he locks the financier back in his cell for the night.

My thanks again to Scientific American *for permission to reprint these twenty columns. As in the two previous book collections, the columns have been expanded, errors corrected and much new material added that was sent to me by readers. I am grateful, also, to my wife for help in proofing; to my editor, Nina Bourne; above all, to that still-growing band of readers, scattered throughout the nation and the world, whose welcome letters have so greatly enriched the material reprinted here.*

MARTIN GARDNER

MARTIN GARDNER'S NEW MATHEMATICAL DIVERSIONS FROM **SCIENTIFIC AMERICAN**

CHAPTER ONE

The Binary System

A red ticket showed between wiper and windshield; I carefully tore it into two, four, eight pieces.

— Vladimir Nabokov, *Lolita*

THE NUMBER SYSTEM now in use throughout the civilized world is a decimal system based on successive powers of 10. The digit at the extreme right of any number stands for a multiple of 10^0, or 1. The second digit from the right indicates a multiple of 10^1; the third digit, a multiple of 10^2, and so on. Thus 777 expresses the sum of $(7 \times 10^0) + (7 \times 10^1) + (7 \times 10^2)$. The widespread use of 10 as a number base is almost certainly due to the fact that we have ten fingers; the very word "digit" reflects this. If Mars is inhabited by humanoids with twelve fingers, it is a good bet that Martian arithmetic uses a notation based on 12.

The simplest of all number systems that make use of the positions of digits is the binary system, based on the powers of 2. Some primitive tribes count in binary fashion, and ancient Chinese mathematicians knew about the system, but it was the great German mathematician Gottfried Wilhelm von Leibniz who seems to have been the first to develop the system in any detail. For Leibniz, it symbolized a deep metaphysical truth. He regarded 0 as an emblem of nonbeing or nothing; 1 as an emblem of being or substance. Both are necessary to the Creator, because a cosmos containing only pure substance would be indistinguishable from the empty cosmos, devoid of sound and fury and signified by 0. Just as in the binary system any integer can be expressed by a suitable placing of 0's and 1's, so the mathematical structure of the entire created world becomes possible, Leibniz believed, as a consequence of the primordial binary split between being and nothingness.

From Leibniz's day until very recently the binary system was little more than a curiosity, of no practical value. Then came the computers! Wires either do or do not carry a current, a switch is on or off, a magnet is north-south or south-north, a flip-flop memory circuit is flipped or flopped. For such reasons enormous speed and accuracy are obtained by constructing computers that can process data coded in binary form. "Alas!" writes Tobias Dantzig in his book *Number, the Language of Science,* "what was once hailed as a monument to monotheism ended in the bowels of a robot."

Many mathematical recreations involve the binary system: the game of Nim, mechanical puzzles such as the Tower of Hanoi and the Rings of Cardan, and countless card tricks and "brainteasers." Here we shall restrict our attention to a familiar set of "mind-reading" cards, and a closely related set of punch-cards with which several remarkable binary feats can be performed.

The construction of the mind-reading cards is made clear in Figure 1. On the left are the binary numbers from 0 through 31. Each digit in a binary number stands for a power of 2, beginning with 2^0 (or 1) at the extreme right, then proceeding leftward to 2^1 (or 2), 2^2, 2^3 and so on. These powers of 2 are shown at the top of the columns. To translate a binary number into its decimal equivalent, simply sum the powers of 2 that are expressed by the positions of the 1's. Thus 10101 represents $16 + 4 + 1$, or 21. To change 21 back to the binary form, a reverse procedure is followed. Divide 21 by 2. The result is 10 with a remainder of 1. This remainder is the first digit on the right of the binary number. Next divide 10 by 2. There is no remainder, so the next binary digit is 0. Then 5 is divided by 2, and so on until the binary number 10101 is completed. In the last step, 2 goes into 1 no times, with a remainder of 1.

The table of binary numbers is converted to a set of mind-reading cards simply by replacing each 1 with the decimal number that corresponds to the binary number in which the 1 occurs. The result is shown at the right side of the illustration. Each column of numbers is copied on a separate card. Hand the five

BINARY NUMBERS

	16	8	4	2	1
0					0
1					1
2				1	0
3				1	1
4			1	0	0
5			1	0	1
6			1	1	0
7			1	1	1
8		1	0	0	0
9		1	0	0	1
10		1	0	1	0
11		1	0	1	1
12		1	1	0	0
13		1	1	0	1
14		1	1	1	0
15		1	1	1	1
16	1	0	0	0	0
17	1	0	0	0	1
18	1	0	0	1	0
19	1	0	0	1	1
20	1	0	1	0	0
21	1	0	1	0	1
22	1	0	1	1	0
23	1	0	1	1	1
24	1	1	0	0	0
25	1	1	0	0	1
26	1	1	0	1	0
27	1	1	0	1	1
28	1	1	1	0	0
29	1	1	1	0	1
30	1	1	1	1	0
31	1	1	1	1	1

MIND-READING CARDS

				1
			2	
			3	3
		4		
		5		5
		6	6	
		7	7	7
	8			
	9			9
	10		10	
	11		11	11
	12	12		
	13	13		13
	14	14	14	
	15	15	15	15
16				
17				17
18			18	
19			19	19
20		20		
21		21		21
22		22	22	
23		23	23	23
24	24			
25	25			25
26	26		26	
27	27		27	27
28	28	28		
29	29	29		29
30	30	30	30	
31	31	31	31	31

FIG. 1

Numbers on a set of mind-reading cards (right) are based on the binary numbers (left).

cards to someone, ask him to think of any number from 0 to 31 inclusive and then to hand you all the cards on which his number appears. You can immediately name the number. To learn it, you have only to add the top numbers of the cards given to you.

How does it work? Each number appears on a unique combination of cards, and this combination is equivalent to the binary notation for that number. When you total the top numbers on the cards, you are simply adding the powers of 2 that are indicated by the 1's in the binary version of the chosen number. The working of the trick can be further disguised by using cards of five different colors. You can then stand across the room and tell your subject to put all the cards bearing his number into a certain pocket and all remaining cards into another pocket. You, of course, must observe this, remembering which power of 2 goes with which color. Another presentation is to put the five (uncolored) cards in a row on the table. Stand across the room and ask the spectator to turn face down those cards that bear his number. Since you arranged the cards with their top numbers in order, you have only to observe which cards are reversed to know which key numbers to add.

The binary basis of punch-card sorting is amusingly dramatized by the set of cards depicted in Figure 2. They can be made easily from a set of 32 file cards. The holes should be a trifle larger than the diameter of a pencil. It is a good plan to cut five holes in one card, then use this card as a stencil for outlining holes on the other cards. If no punching device is available, the cutting of the holes with scissors can be speeded by holding three cards as one and cutting them simultaneously. The cut-off corners make it easy to keep the cards properly oriented. After five holes have been made along the top of each card, the margin is trimmed away above certain holes as shown in the illustration. These notched holes correspond to the digit 1; the remaining holes correspond to 0. Each card carries in this way the equivalent of a binary number. The numbers run from 0 through 31, but in the illustration the cards are randomly arranged. Three unusual stunts can be performed with these cards. They may be troublesome to make, but everyone in the family will enjoy playing with them.

The first stunt consists of quickly sorting the cards so that

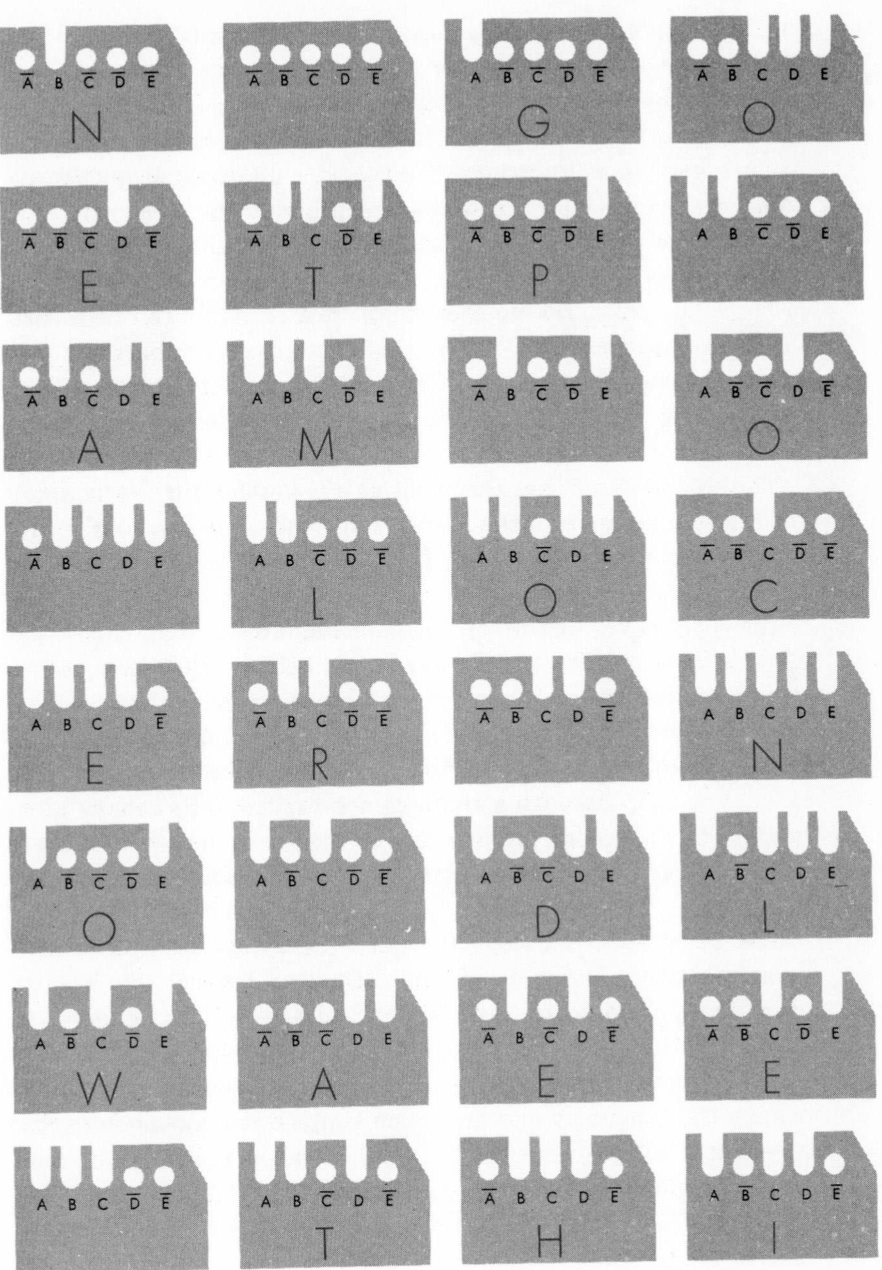

FIG. 2
A set of punch-cards that will unscramble a message, guess a selected number and solve logic problems.

their numbers are in serial order. Mix the cards any way you please, then square them like a deck of playing cards. Insert a pencil through hole E and lift up an inch or so. Half the cards will cling to the pencil, and half will be left behind. Give the pencil some vigorous shakes to make sure all cards drop that are supposed to drop, then raise the pencil higher until the cards are separated into two halves. Slide the packet off the pencil and put it in *front* of the other cards. Repeat this procedure with each of the other holes, taking them right to left. After the fifth sorting, it may surprise you to find that the binary numbers are now in serial order, beginning with 0 on the card facing you. Flip through the cards and read the Christmas message printed on them!

The second stunt uses the cards as a computer for determining the selected number on the set of mind-reading cards. Begin with the punch-cards in any order. Insert the pencil in hole E and ask if the chosen number appears on the card with a top number of 1. If the reply is yes, lift up on the pencil and discard all cards that cling to it. If no, discard all cards left behind. You now have a packet of sixteen cards. Ask if the number is on the card with a top number of 2, then repeat the procedure with the pencil in hole D. Continue in this manner with the remaining cards and holes. You will end with a single punch-card, and its binary number will be the chosen number. If you like, print decimal numbers on all the cards so that you will not have to translate the binary numbers.

The third stunt employs the cards as a logic computer in a manner first proposed by William Stanley Jevons, the English economist and logician. Jevons' "logical abacus," as he called it, used flat pieces of wood with steel pins at the back so that they could be lifted from a ledge, but the punch-cards operate in exactly the same way and are much simpler to make. Jevons also invented a complex mechanical device, called the "logic piano," which operated on the same principles, but the punch-cards will do all that his piano could do. In fact, they will do more, because the piano took care of only four terms, and the cards take care of five.

The five terms A, B, C, D and E are represented by the five holes, which in turn represent binary digits. Each 1 (or notched hole) corresponds to a true term; each 0, to a false term. A line over the top of a letter indicates that the term is false; otherwise it is true. Each card is a unique combination of true and false terms, and since the 32 cards exhaust all possible combinations, they are the equivalent of what is called a "truth table" for the five terms. The operation of the cards is best explained by showing how they can be used for solving a problem in two-valued logic.

The following puzzle appears in *More Problematical Recreations,* a booklet issued recently by Litton Industries of Beverly Hills, California. "If Sara shouldn't, then Wanda would. It is impossible that the statements: 'Sara should,' and 'Camille couldn't,' can both be true at the same time. If Wanda would, then Sara should and Camille could. Therefore Camille could. Is the conclusion valid?"

To solve this problem, start with the punch-cards in any order. Only three terms are involved, so we shall be concerned with only the A, B and C holes.

A = Sara should
$\overline{A}$ = Sara shouldn't
B = Wanda would
$\overline{B}$ = Wanda wouldn't
C = Camille could
$\overline{C}$ = Camille couldn't

The problem has three premises. The first—"If Sara shouldn't, then Wanda would"—tells us that the combination of $\overline{A}$ and $\overline{B}$ is not permitted, so we must eliminate all cards bearing this combination. It is done as follows. Insert the pencil in A and lift. All cards on the pencil bear $\overline{A}$. Hold them as a group, remove the pencil, put it in B and lift again. The pencil will raise all cards bearing both $\overline{A}$ and $\overline{B}$, the invalid combination, so these cards are discarded. All remaining cards are assembled into a pack once more (the order does not matter), and we are ready for the second premise.

Premise two is that "Sara should" and "Camille couldn't" cannot both be true. In other words, we cannot permit the combination $A\overline{C}$. Insert the pencil in A and lift up all cards bearing $\overline{A}$.

These are *not* the cards we want, so we place them temporarily aside and continue with the A group that remains. Insert the pencil in C and raise the $\overline{C}$ cards. These bear the invalid combination $A\overline{C}$, so they are permanently discarded. Assemble the remaining cards once more.

The last premise tells us that if Wanda would, then Sara should and Camille could. A bit of reflection will show that this eliminates two combinations: $\overline{A}B$ and $B\overline{C}$. Put the pencil in hole A, lift, and continue working with the lifted cards. Insert pencil in B; lift. No cards cling to the pencil. This means that the two previous premises have already eliminated the combination $\overline{A}\overline{B}$. Since the cards all bear $\overline{A}B$ (an invalid combination), this entire packet is permanently discarded. The only remaining task is to eliminate $B\overline{C}$ from the remaining cards. The pencil in B will lift out the $\overline{B}$ cards, which are placed temporarily aside. When the pencil is put in C of the cards that remain, no cards can be lifted, indicating that the invalid combination of $B\overline{C}$ has already been ruled out by previous steps.

We are thus left with eight cards, each bearing a combination of truth values for A, B and C that is consistent with all three premises. These combinations are the valid lines of a truth table for the combined premises. Inspection of the cards reveals that C is true on all eight, so it is correct to conclude that Camille could. Other conclusions can also be deduced from the premises. We can, for example, assert that Sara should. But the interesting question of whether Wanda would or wouldn't remains, at least in the light of available knowledge, an inscrutable binary mystery.

For those who would like another problem to feed the cards, here is an easy one. In a suburban home live Abner, his wife Beryl and their three children, Cleo, Dale and Ellsworth. The time is 8 P.M. on a winter evening.

1. If Abner is watching television, so is his wife.
2. Either Dale or Ellsworth, or both of them, are watching television.
3. Either Beryl or Cleo, but not both, is watching television.
4. Dale and Cleo are either both watching or both not watching television.
5. If Ellsworth is watching television, then Abner and Dale are also watching.

Who is watching television and who is not?

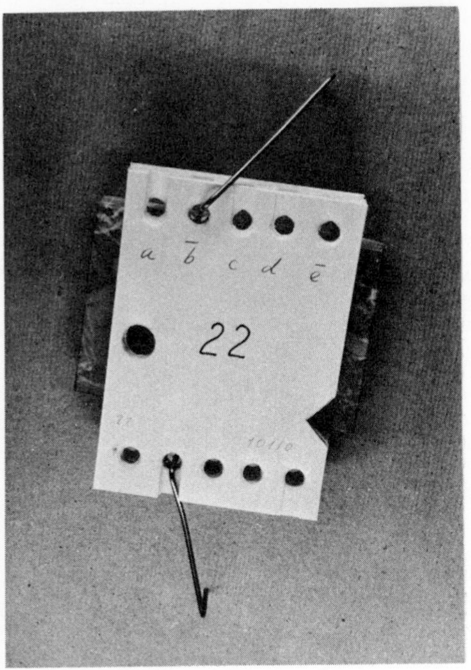

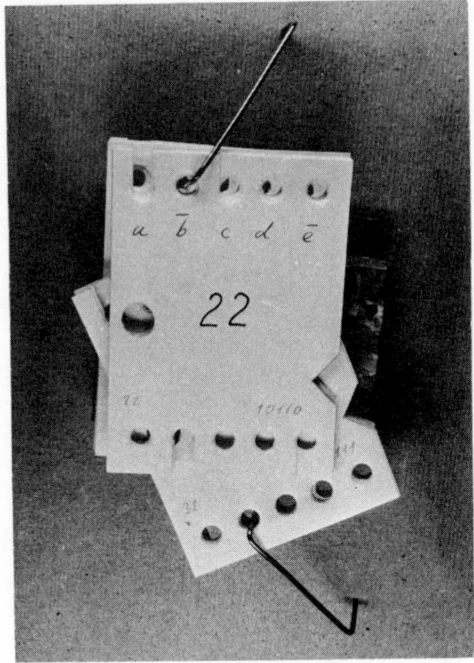

FIG. 3
A complementary row of holes at bottom of cards permits errorless sorting.

ADDENDUM

EDWARD B. GROSSMAN, New York City, wrote to say that a variety of commercial cards for binary filing and sorting are now available in large stationery supply stores. Holes are preperforated and one can buy special punches for making the slots. The holes are too small to take pencils, but one can use knitting needles, Q-Tip sticks, opened-out paper clips or the sorting rods that come with some makes of cards.

Giuseppe Aprile, a professor of engineering at the University of Palermo, Italy, sent the two photographs shown in Figure 3. Quick, errorless separation of the cards is made possible by a complementary row of holes and notches at the bottom edge of each card. Pins through complementary holes in the bottom row anchor the set of cards that remains when pins through the top holes remove a set of cards.

ANSWERS

THE LOGIC PROBLEM can be solved with the punch-cards as follows: Let A, B, C, D and E stand for Abner, Beryl, Cleo, Dale and Ellsworth. A term is true if the person is watching television; otherwise it is false. Premise 1 eliminates all cards bearing $A\overline{B}$; premise 2 eliminates $\overline{D}\overline{E}$; premise 3 eliminates BC and $\overline{B}\overline{C}$; premise 4 eliminates $\overline{C}D$ and $C\overline{D}$; premise 5 eliminates $\overline{A}E$ and $\overline{D}E$. Only one card remains, bearing the combination $\overline{A}\overline{B}CD\overline{E}$. We conclude that Cleo and Dale are watching television, and that the others are not.

CHAPTER TWO

□

Group Theory and Braids

THE CONCEPT OF "GROUP," one of the great unifying ideas of modern algebra and an indispensable tool in physics, has been likened by James R. Newman to the grin of the Cheshire cat. The body of the cat (algebra as traditionally taught) vanishes, leaving only an abstract grin. A grin implies something amusing. Perhaps we can make group theory less mysterious if we do not take it too seriously.

Three computer programmers — Ames, Baker and Coombs — wish to decide who pays for the beer. Of course they can flip pennies, but they prefer a random decision based on the following network-tracing game. Three vertical lines are drawn on a sheet of paper. One programmer, holding the paper so that his friends cannot see what he is doing, randomly labels the lines A, B and C [*see Fig. 4, left*]. He folds back the top of the sheet to conceal these letters. A second player now draws a series of random horizontal lines — call them shuttles — each connecting two of the vertical lines [*see second illustration of figure*]. The third player adds a few more shuttles, then marks an X at the bottom of one of the vertical lines [*see third illustration*].

The paper is unfolded. Ames puts his finger on the top of line A and traces it downward. When he reaches the end of a shuttle (ignoring shuttles whose centers he may cross), he turns, follows the shuttle to its other end, turns again and continues downward

FIG. 4
The network-tracing game.

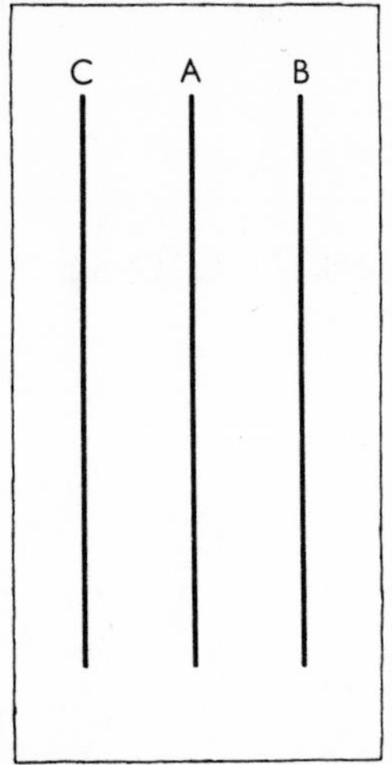

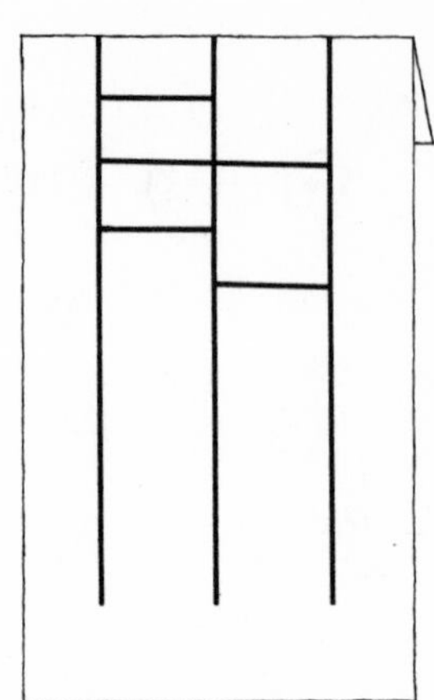

until he reaches the end of another shuttle. He keeps doing this until he reaches the bottom. His path [*traced by the broken line in the fourth illustration*] does not end on the X, so he does not have to buy the drinks. Baker and Coombs now trace their lines in similar fashion. Baker's path ends on the X, so he picks up the tab. For any number of vertical lines, and regardless of how the shuttles are drawn, each player will always end on a different line.

A closer look at this game discloses that it is based on one of the simplest of groups, the so-called permutation group for three symbols. What, precisely, is a group? It is an abstract structure involving a set of undefined elements (a, b, c . . .) and a single undefined binary operation (here symbolized by $\circ$) that pairs one element with another to produce a third. The structure is not a group unless it has the following four traits:

1. When two elements of the set are combined by the operation, the result is another element in the same set. This is called "closure."

2. The operation obeys the "associative law": $(a \circ b) \circ c = a \circ (b \circ c)$

3. There is one element e (called the "identity") such that $a \circ e = e \circ a = a$

4. For every element a there is an inverse element a' such that $a \circ a' = a' \circ a = e$

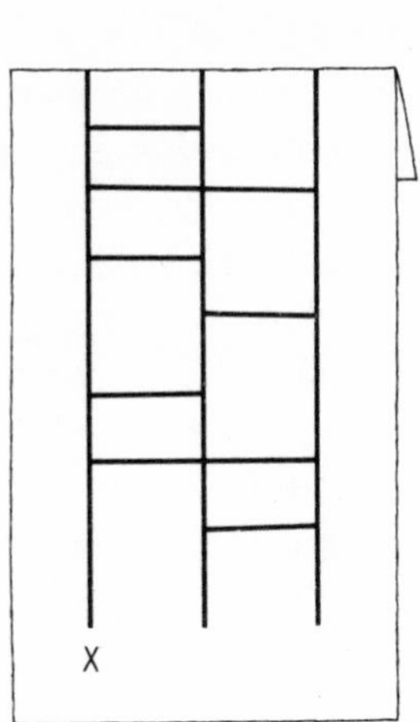

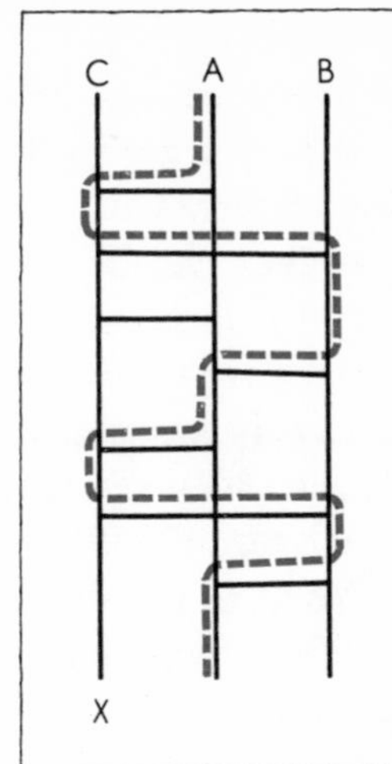

If in addition to these four traits the operation also obeys the commutative law ($a \circ b = b \circ a$), the group is called a commutative or Abelian group.

The most familiar example of a group is provided by the integers (positive, negative and zero) with respect to the operation of addition. It has closure (any integer plus any integer is an integer). It is associative (adding 2 to 3 and then adding 4 is the same as adding 2 to the sum of 3 and 4). The identity is 0 and the inverse of a positive integer is the negative of that integer. It is an Abelian group (2 plus 3 is the same as 3 plus 2). The integers do *not* form a group with respect to division: 5 divided by 2 is $2\frac{1}{2}$, which is not an element in the set.

Let us see how the network game exhibits group structure. Figure 5 depicts the six basic "transformations" that are the elements of our finite group. Transformation p switches the paths of A and B so that the three paths end in the order BAC. Transformations q, r, s and t give other permutations. Transformation e is not really a change, but mathematicians call it a "transformation" anyway, in the same sense that a null or empty class is called a class. It consists of drawing no shuttles at all; it is the "identity" change that doesn't really change anything. These six elements correspond to the six different ways in which three symbols can be permuted. Our group operation, symbolized by $\circ$, is simply that of following one transformation with another; that is, of adding shuttles.

A quick check reveals that we have here a structure with all the properties of a group. It has closure, because no matter how we pair the elements we always get a permutation in the order of the paths that can be achieved by one element alone. For example, $p \circ t = r$, because p followed by t has exactly the same effect on the path order as applying r alone. The operation of adding shuttles is clearly associative. Adding no shuttles is the identity. Elements p, q and r are their own inverses, and s and t are inverses of each other. (When an element and its inverse are combined, the result is the same as drawing no shuttles at all.) It is not an Abelian group (*e.g.*, p followed by q is not the same as q followed by p).

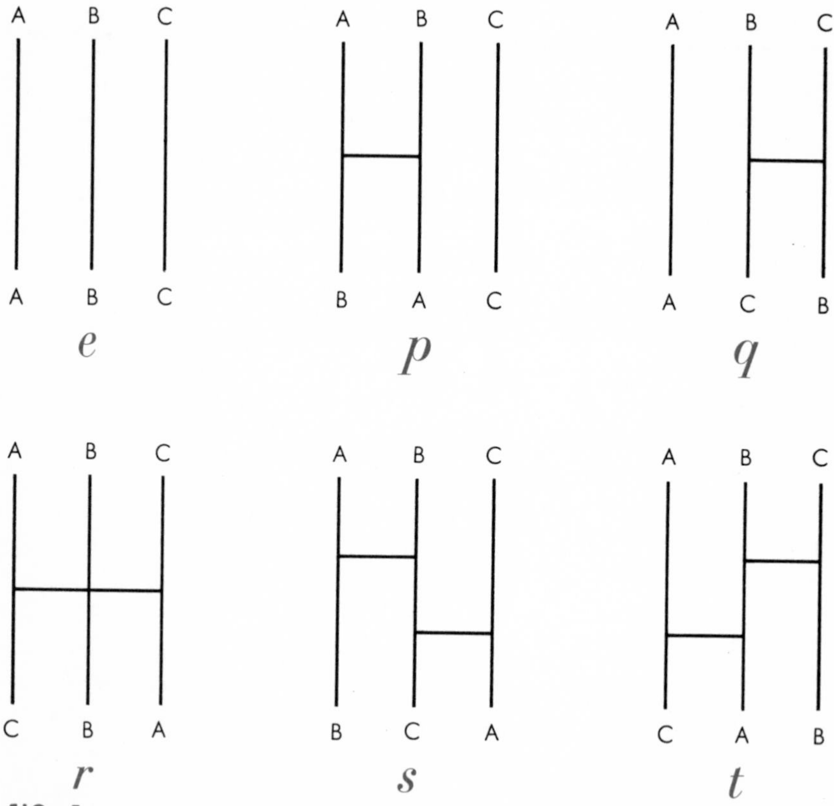

FIG. 5
The six elements of the network-game group.

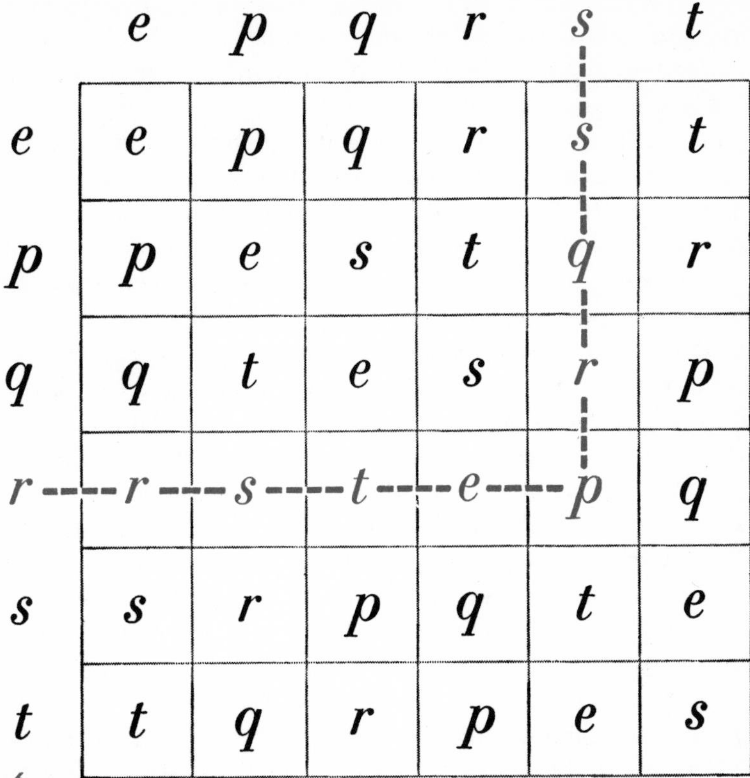

	e	p	q	r	s	t
e	e	p	q	r	s	t
p	p	e	s	t	q	r
q	q	t	e	s	r	p
r	r	s	t	e	p	q
s	s	r	p	q	t	e
t	t	q	r	p	e	s

FIG. 6
Results of pairing elements in the network-game group.

The table in Figure 6 provides a complete description of this group's structure. What is the result of following r with s? We find r on the left side of the table and s at the top. The intersection of column and row is the cell labeled p. In other words, shuttle pattern r followed by shuttle pattern s has the same effect on path order as pattern p. This is a very elementary group that turns up in many places. For example, if we label the corners of an equilateral triangle, then rotate and reflect the triangle so that it always occupies the same position on the plane, we find that there are only six basic transformations possible. These transformations have the same structure as the group just described.

It is not necessary to go into group theory to see intuitively that the network game will never permit two players to end their paths on the same vertical line. Simply think of the three lines as three ropes. Each shuttle has the same effect on path order as crossing two ropes, as though forming a braid. Obviously no matter how you make the braid or how long it is, there will always be three separate lower ends.

Let us imagine that we are braiding three strands of a girl's hair. We can record successive permutations of strands by means of the network diagram, but it will not show how the strands pass over and under one another. If we take into account this complicating topological factor, is it still possible to call on group theory for a description of what we are doing? The answer is yes, and Emil Artin, a distinguished German mathematician who died in 1962, was the first to prove it. In his elegant theory of braids the elements of the group are "weaving patterns" (infinite in number), and the operation consists, as in the network game, of following one pattern with another. As before, the identity element is a pattern of straight strands — the result of doing nothing. The inverse of a weaving pattern is its mirror image. Figure 7 shows a sample pattern followed by its inverse. Group theory tells us that when an element is added to its inverse, the result is the identity. Sure enough, the two weaving patterns combined prove to be topologically equivalent to the identity. A tug on the end of the braid in the illustration and all strands pull out straight. (Many magic tricks with rope, known in the trade as releases, are based on this interesting property of groups. For a good one, see Chapter 7 of *The 2nd Scientific American Book of Mathematical Puzzles & Diversions*.) Artin's theory of braids not only provided for the first time a system that classified all types of braids; it also furnished a method by which one could determine whether two weaving patterns, no matter how complex, were or were not topologically equivalent.

Braid theory is involved in an unusual game devised by the Danish poet, writer and mathematician Piet Hein. Cut a piece of heavy cardboard into the coat-of-arms shape depicted in Figure 8. This will be called the plaque. Its two sides must be easily distinguished, so color one side or mark it with an X as shown. Punch three holes at the square end. A two-foot length of heavy but flexible cord (sash cord is excellent) is knotted to each hole. The other ends of the three strands are attached to some fixed object like the back of a chair.

FIG. 7
Braid A is the mirror image of A′.

You will find that the plaque can be given complete rotations in six different ways to form six different braids. It can be rotated sidewise to the right or to the left; it can be rotated forward or backward between strands A and B; it can be rotated forward or backward between strands B and C. The second illustration of Figure 8 shows the braid obtained by a forward rotation through B and C. The question arises: Is it possible to untangle this braid by weaving the plaque in and out through the strands, keeping it horizontal at all times, X-side up, and always pointing toward you? The answer is no. But if you give the plaque a second rotation, in any of the six different ways, the result is a braid that *can* be untangled by weaving the plaque without rotating it.

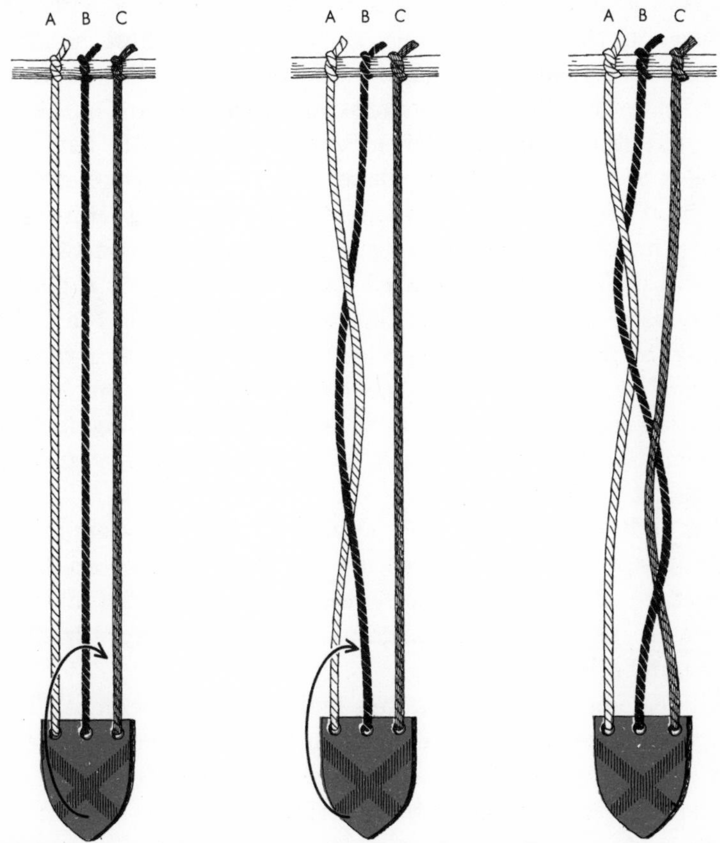

FIG. 8
Rotation at left produces braid in center; rotation in center, braid at right.

To make this clear, assume that the second rotation is forward between A and B, creating the braid shown in the third illustration of Figure 8. To remove this braid without rotating the plaque, first raise C at the spot marked Y and pass the plaque under it from right to left. Pull the strings taut. Next raise A at the spot marked Z and pass the plaque under it from left to right. The result is that the cords pull straight.

The following surprising theorem holds for any number of strands above two. All braids produced by an *even* number of rotations (each rotation may be in any direction whatever) can always be untangled by weaving the plaque without rotating it; braids produced by an *odd* number of full rotations can never be untangled.

It was at a meeting in Niels Bohr's Institute for Theoretical Physics, in the early thirties, that Piet Hein first heard this theorem discussed by Paul Ehrenfest in connection with a problem in quantum theory. A demonstration was worked out, by Piet Hein and others, in which Mrs. Bohr's scissors were fastened to the back of a chair with strands of cord. It later occurred to Piet Hein that the rotating body and the surrounding universe entered symmetrically into the problem and therefore that a symmetrical model could be created simply by attaching a plaque to *both* ends of the cord. With this model two persons can play a topological game. Each holds a plaque, and the three strands are stretched straight between the two plaques. The players take turns, one forming a braid and the other untangling it, timing the operation to see how long it takes. The player who untangles the fastest is the winner.

The odd-even theorem also applies to this two-person game. Beginners should limit themselves to two-rotation braids, then proceed to higher even-order braids as they develop skill. Piet Hein calls his game Tangloids, and it has been played in Europe for a number of years.

Why do odd and even rotations make such a difference? This is a puzzling question that is difficult to answer without going more deeply into group theory. A hint is supplied by the fact that two rotations in exactly opposite directions naturally amount to no rotation. And if two rotations are almost opposite, prevented from being so only by the way certain strings pass around the plaque, then the tangle can be untangled by moving those same

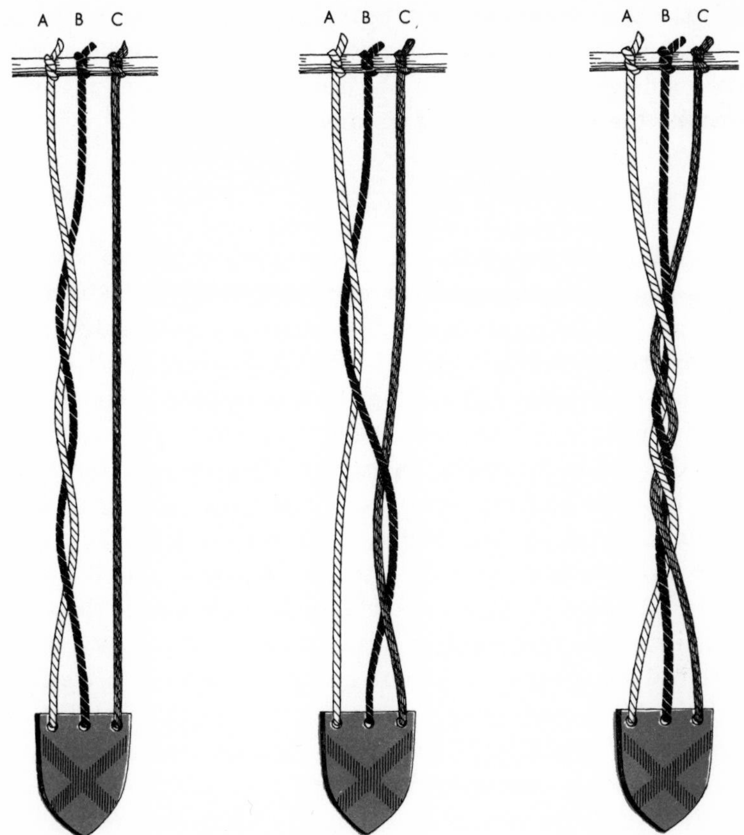

FIG. 9
Three problems of braid disentanglement.

strings back around the plaque. M. H. A. Newman, in an article published in a London mathematical journal in 1942, says that P. A. M. Dirac, the noted University of Cambridge physicist, has for many years used the solitaire form of this game as a model "to illustrate the fact that the fundamental group of the group of rotations in 3-space has a single generator of the period 2." Newman then draws on Artin's braid theory to prove that the cords cannot be untangled when the number of rotations is odd.

You will find it a fascinating pastime to form braids by randomly rotating the plaque an even number of times, then seeing how quickly you can untangle the cords. Three simple braids, each formed by two rotations, are shown in Figure 9. The braid on the left is formed by rotating the plaque forward twice through B and C; the braid in center, by rotating the plaque forward through B and C and then backward through A and B; the braid at right, by two sidewise rotations to the right. Readers are invited to determine the best method of untangling each braid.

ADDENDUM

IN CONSTRUCTING the device used for playing Piet Hein's game of Tangloids, plaques cut from flat pieces of wood or plastic are, of course, preferable to cardboard. Instead of three separate strands, Piet Hein recommends using one long single cord. Start it at the first hole of one plaque (knotting the end to keep it from sliding out of the hole), run it through the first hole of the second plaque, across the plaque, through its middle hole, then to the middle hole of the first plaque, across to its third hole, and back to the third hole of the second plaque, knotting the end after it has passed through this last hole. Because the cord can slide freely through the holes, it is easier to manipulate the device than when it has three separate strands. One reader wrote to say that he had joined his plaques with three strands of *elastic* cord, and found that this also made the manipulations much easier. The game can obviously be elaborated by adding more strands, but three seems to be complicated enough.

It takes only a glance at the table in Figure 6 to see that the group it depicts is not Abelian (commutative). Tables for Abelian groups are symmetrical along an axis running from upper left to lower right corner. That is, the triangular sections on either side of this diagonal are mirror images of each other.

If the network game is played by four players instead of three, its group is the permutation group for four symbols. This is not, however, identical with the group that describes the rotations and reflections of a square, because certain permutations of the corners of a square are not obtainable by rotating and reflecting it. The square transformations are a "subgroup" of the permutation group for four symbols. All finite groups (groups with a finite number of elements) are either permutation groups or subgroups of permutation groups.

In Artin's 1947 paper on braid theory (see the Bibliography) he gives a method of reducing any braid to "normal form." This involves pulling the first strand completely straight. The second strand is then pulled straight except for its loops around strand 1. Strand 3 is then pulled straight except for its loops around strands 1 and 2, and so on for the remaining strands. "Although it has been proved that every braid can be deformed into a similar normal form," Artin says, "the writer is convinced that any attempt to carry this out on a living person would only lead to violent protests and discrimination against mathematics."

In a brief letter from Dirac that I received too late to mention in the column on braids, he said that he had first thought of the string problem about 1929, and had used it many times since to illustrate that two rotations of a body about an axis can be continuously deformed, through a set of motions each of which ends with the original position, into no motion at all. "It is a consequence," he wrote, "of this property of rotations that a spinning body can have half a quantum of angular momentum, but cannot have any other fraction of a quantum."

ANSWERS

THE THREE braid problems are solved as follows: (1) Pass the plaque under strand C from right to left, then under strands A and B from left to right. (2) Pass the plaque under the center of strand B from left to right. (3) Pass the plaque, left to right, under all strands.

CHAPTER THREE

□

Eight Problems

1. ACUTE DISSECTION

GIVEN A TRIANGLE with one obtuse angle, is it possible to cut the triangle into smaller triangles, all of them acute? (An acute triangle is a triangle with three acute angles. A right angle is of course neither acute nor obtuse.) If this cannot be done, give a proof of impossibility. If it can be done, what is the smallest number of acute triangles into which any obtuse triangle can be dissected?

Figure 10 shows a typical attempt that leads nowhere. The triangle has been divided into three acute triangles, but the fourth is obtuse, so nothing has been gained by the preceding cuts.

The problem (which came to me by way of Mel Stover of Winnipeg) is amusing because even the best mathematician is likely to be led astray by it and come to a wrong conclusion. My pleasure

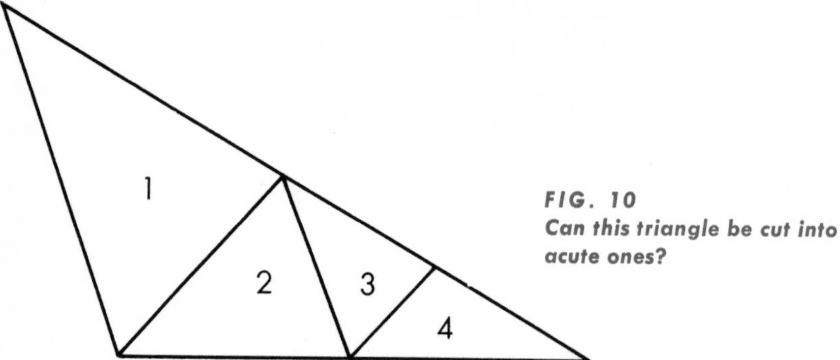

FIG. 10
Can this triangle be cut into acute ones?

in working on it led me to ask myself a related question: "What is the smallest number of acute triangles into which a *square* can be dissected?" For days I was convinced that nine was the answer; then suddenly I saw how to reduce it to eight. I wonder how many readers can discover an eight-triangle solution, or perhaps an even better one. I am unable to prove that eight is the minimum, though I strongly suspect that it is.

2. HOW LONG IS A "LUNAR"?

IN H. G. WELLS'S NOVEL *The First Men in the Moon* our natural satellite is found to be inhabited by intelligent insect creatures who live in caverns below the surface. These creatures, let us assume, have a unit of distance that we shall call a "lunar." It was adopted because the moon's surface area, if expressed in square lunars, exactly equals the moon's volume in cubic lunars. The moon's diameter is 2,160 miles. How many miles long is a lunar?

3. THE GAME OF GOOGOL

IN 1958 JOHN H. FOX, JR., of the Minneapolis-Honeywell Regulator Company, and L. Gerald Marnie, of the Massachusetts Institute of Technology, devised an unusual betting game which they call Googol. It is played as follows: Ask someone to take as many slips of paper as he pleases, and on each slip write a different positive number. The numbers may range from small fractions of 1 to a number the size of a "googol" (1 followed by a hundred 0's) or even larger. These slips are turned face down and shuffled over the top of a table. One at a time you turn the slips face up. The aim is to stop turning when you come to the number that you guess to be the largest of the series. You cannot go back and pick a previously turned slip. If you turn over all the slips, then of course you must pick the last one turned.

Most people will suppose the odds against your finding the highest number to be at least five to one. Actually if you adopt the best strategy, your chances are a little better than one in three. Two questions arise. First, what is the best strategy? (Note that this is not the same as asking for a strategy that will maximize the *value* of the selected number.) Second, if you follow this strategy, how can you calculate your chances of winning?

When there are only two slips, your chance of winning is obviously 1/2, regardless of which slip you pick. As the slips increase in number, the probability of winning (assuming that you use the best strategy) decreases, but the curve flattens quickly, and there is very little change beyond ten slips. The probability never drops below 1/3. Many players will suppose that they can make the task more difficult by choosing very large numbers, but a little reflection will show that the sizes of the numbers are irrelevant. It is only necessary that the slips bear numbers that can be arranged in increasing order.

The game has many interesting applications. For example, a girl decides to marry before the end of the year. She estimates that she will meet ten men who can be persuaded to propose, but once she has rejected a proposal, the man will not try again. What strategy should she follow to maximize her chances of accepting the top man of the ten, and what is the probability that she will succeed?

The strategy consists of rejecting a certain number of slips of paper (or proposals), then picking the next number that exceeds the highest number among the rejected slips. What is needed is a formula for determining how many slips to reject, depending on the total number of slips.

4. MARCHING CADETS AND A TROTTING DOG

A SQUARE FORMATION of Army cadets, 50 feet on the side, is marching forward at a constant pace [*see Fig. 11*]. The company mascot, a small terrier, starts at the center of the rear rank [*position A in the illustration*], trots forward in a straight line to

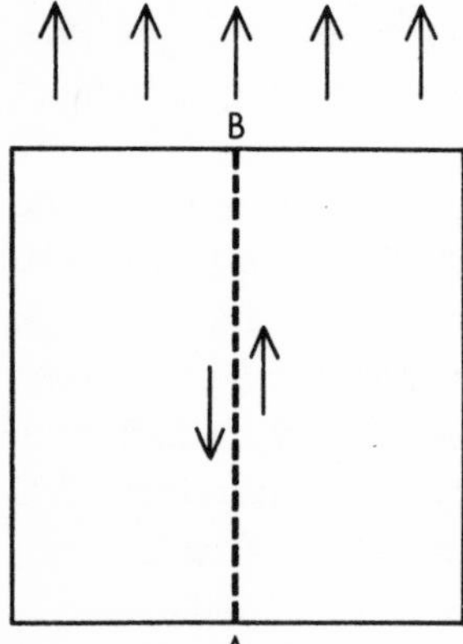

FIG. 11
How far does the dog trot?

the center of the front rank [*position B*], then trots back again in a straight line to the center of the rear. At the instant he returns to position A, the cadets have advanced exactly 50 feet. Assuming that the dog trots at a constant speed and loses no time in turning, how many feet does he travel?

If you solve this problem, which calls for no more than a knowledge of elementary algebra, you may wish to tackle a much more difficult version proposed by the famous puzzlist Sam Loyd. (See *Mathematical Puzzles of Sam Loyd,* Vol. 2, Dover paperback, 1960, page 103.) Instead of moving forward and back through the marching cadets, the mascot trots with constant speed around the *outside* of the square, keeping as close as possible to the square at all times. (For the problem we assume that he trots along the perimeter of the square.) As before, the formation has marched 50 feet by the time the dog returns to point A. How long is the dog's path?

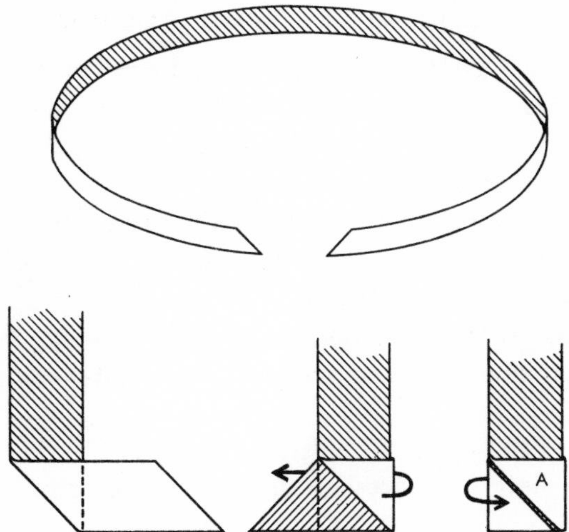

FIG. 12
Barr's belt (top) *and an unsatisfactory way to fold it* (bottom).

5. BARR'S BELT

STEPHEN BARR of Woodstock, New York, says that his dressing gown has a long cloth belt, the ends of which are cut at 45-degree angles as shown in Figure 12. When he packs the belt for a trip, he likes to roll it up as neatly as possible, beginning at one end, but the slanting ends disturb his sense of symmetry. On the other hand, if he folds over an end to make it square off, then the uneven thicknesses of cloth put lumps in the roll. He experimented with more complicated folds, but try as he would, he could not achieve a rectangle of uniform thickness. For example, the fold shown in the illustration produces a rectangle with three thicknesses in section A and two in section B.

"Nothing is perfect," says one of the Philosophers in James Stephens' *The Crock of Gold.* "There are lumps in it." Nonetheless, Barr finally managed to fold his belt so that each end was straight across and part of a rectangle of uniform thickness throughout. The belt could then be folded into a neat roll, free of lumps. How did Barr fold his belt? A long strip of paper, properly cut at the ends, can be used for working on the problem.

6. WHITE, BLACK AND BROWN

PROFESSOR MERLE WHITE of the mathematics department, Professor Leslie Black of philosophy, and Jean Brown, a young stenographer who worked in the university's office of admissions, were lunching together.

"Isn't it remarkable," observed the lady, "that our last names are Black, Brown and White and that one of us has black hair, one brown hair and one white."

"It is indeed," replied the person with black hair, "and have you noticed that not one of us has hair that matches his or her name?"

"By golly, you're right!" exclaimed Professor White.

If the lady's hair isn't brown, what is the color of Professor Black's hair?

7. THE PLANE IN THE WIND

AN AIRPLANE FLIES in a straight line from airport A to airport B, then back in a straight line from B to A. It travels with a constant engine speed and there is no wind. Will its travel time for the same round trip be greater, less or the same if, throughout both flights, at the same engine speed, a constant wind blows from A to B?

8. WHAT PRICE PETS?

THE OWNER of a pet shop bought a certain number of hamsters and half that many pairs of parakeets. He paid $2 each for the hamsters and $1 for each parakeet. On every pet he placed a retail price that was an advance of 10 per cent over what he paid for it.

After all but seven of the creatures had been sold, the owner found that he had taken in for them an amount of money exactly equal to what he had originally paid for all of them. His potential profit, therefore, was represented by the combined retail value of the seven remaining animals. What was this value?

ANSWERS

1. A number of readers sent "proofs" that an obtuse triangle cannot be dissected into acute triangles, but of course it can. Figure 13 shows a seven-piece pattern that applies to any obtuse triangle.

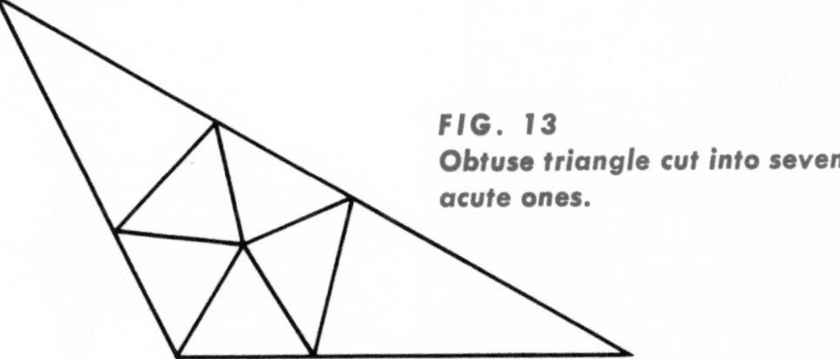

FIG. 13
Obtuse triangle cut into seven acute ones.

It is easy to see that seven is minimal. The obtuse angle must be divided by a line. This line cannot go all the way to the other side, for then it would form another obtuse triangle, which in turn would have to be dissected, consequently the pattern for the large triangle would not be minimal. The line dividing the obtuse angle must, therefore, terminate at a point *inside* the triangle. At this vertex, at least five lines must meet, otherwise the angles at this vertex would not all be acute. This creates the inner pentagon of five triangles, making a total of seven triangles in all. Wallace Manheimer, a Brooklyn high school student at the time, gave this proof as his solution to problem E1406 in *American Mathematical Monthly*, November 1960, page 923. He also showed how to construct the pattern for any obtuse triangle.

The question arises: Can any obtuse triangle be dissected into seven acute *isosceles* triangles? The answer is no. Verner E. Hoggatt, Jr., and Russ Denman (*American Mathematical Monthly,* November 1961, pages 912–913) proved that eight such triangles are sufficient for all obtuse triangles, and Free Jamison (*ibid.,* June-July 1962, pages 550–552) proved that eight are also necessary. These articles can be consulted for details as to conditions under which less than eight-piece patterns are possible. A right triangle and an acute nonisosceles triangle can each be cut into nine acute isosceles triangles, and an acute isosceles triangle can be cut into four congruent acute isosceles triangles similar to the original.

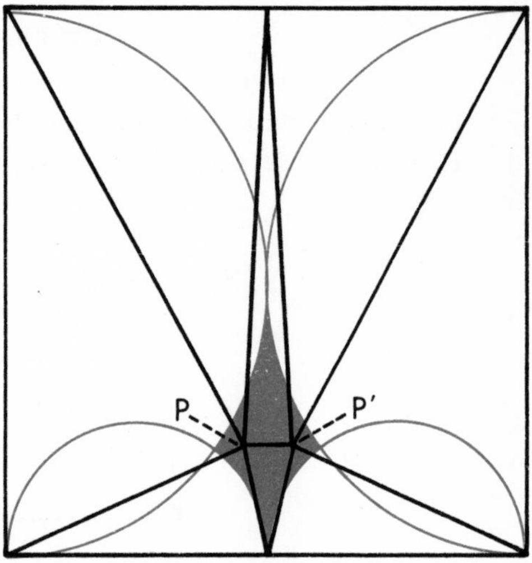

FIG. 14
Square cut into eight acute triangles.

A square can be cut into eight acute triangles as shown in Figure 14. If the dissection has bilateral symmetry, points P and P′ must lie within the shaded area determined by the four semicircles. Donald L. Vanderpool pointed out in a letter that asymmetric distortions of the pattern are possible with point P anywhere outside the shaded area provided it remains outside the two large semicircles.

About 25 readers sent proofs, with varying degrees of formality, that the eight-piece dissection is minimal. One, by Harry Lindgren, appeared in *Australian Mathematics Teacher,* Vol. 18, pages 14–15, 1962. His proof also shows that the pattern, aside from the shifting of points P and P′ as noted above, is unique.

H. S. M. Coxeter pointed out the surprising fact that for any rectangle, even though its sides differ in length by an arbitrarily small amount, the line segment PP′ can be shifted to the center to give the pattern both horizontal and vertical symmetry.

Two unanswered questions: Terence C. Terman divided the square into eleven acute isosceles triangles and wondered if this is minimal. Alan Sutcliffe asked if there is a quadrilateral that cannot be divided into eight or fewer acute triangles.

Figure 15 shows how the pentagram (regular five-pointed star) and the Greek cross can each be dissected into the smallest possible number of acute triangles.

2. The volume of a sphere is $4\pi/3$ times the cube of the radius. Its surface is 4π times the square of the radius. If we express the moon's radius in "lunars" and assume that its surface in square lunars equals its volume in cubic lunars, we can determine the length of the radius simply by equating the two formulas and solving for the value of the radius. Pi cancels out on both sides, and we find that the radius is three lunars. The moon's radius is 1,080 miles, so a lunar must be 360 miles.

3. Regardless of the number of slips involved in the game of Googol, the probability of picking the slip with the largest number (assuming that the best strategy is used) never drops below .367879. This is the reciprocal of e, and the limit of the probability of winning as the number of slips approaches infinity.

If there are ten slips (a convenient number to use in playing the game), the probability of picking the top number is .398. The strategy is to turn three slips, note the largest number among them, then pick the next slip that exceeds this number. In the long run you stand to win about two out of every five games.

What follows is a compressed account of a complete analysis of the game by Leo Moser and J. R. Pounder of the University of Alberta. Let n be the number of slips and p the number rejected before picking a number larger than any on the p slips. Number the slips serially from 1 to n. Let $k + 1$ be the number of the slip bearing the largest number. The top number will not be chosen unless k is greater than p (otherwise the number will be rejected among the first p slips), and then only if the highest number from 1 to k is also the highest number from 1 to p (otherwise *this* number will be chosen before the top number is reached). The prob-

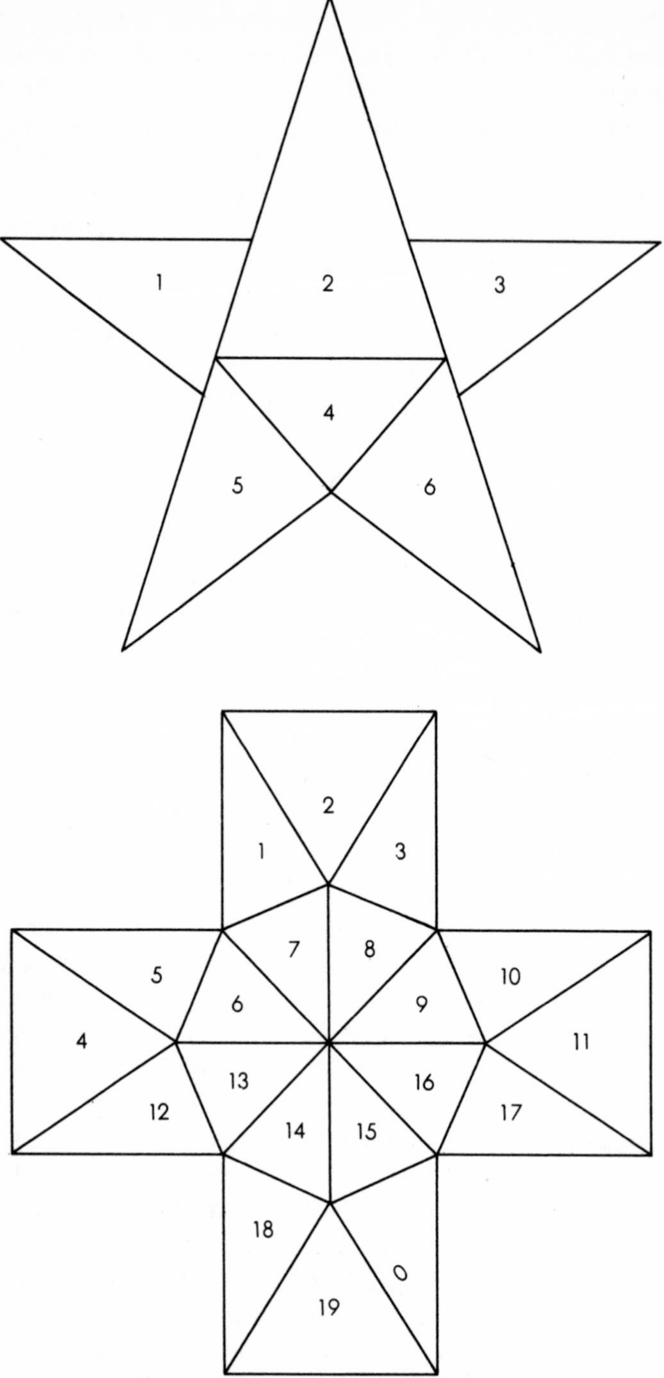

FIG. 15
Minimum dissections for the pentagram and Greek cross.

ability of finding the top number in case it is on the $k + 1$ slip is p/k, and the probability that the top number actually is on the $k + 1$ slip is $1/n$. Since the largest number can be on only one slip, we can write the following formula for the probability of finding it:

$$\frac{p}{n}\left(\frac{1}{p} + \frac{1}{p+1} + \frac{1}{p+2} \cdots\cdot + \frac{1}{n-1}\right)$$

Given a value for n (the number of slips) we can determine p (the number to reject) by picking a value for p that gives the greatest value to the above expression. As n approaches infinity, p/n approaches $1/e$, so a good estimate of p is simply the nearest integer to n/e. The general strategy, therefore, when the game is played with n slips, is to let n/e numbers go by, then pick the next number larger than the largest number on the n/e slips passed up.

This assumes, of course, that a player has no knowledge of the range of the numbers on the slips and therefore no basis for knowing whether a single number is high or low within the range. If one *has* such knowledge, the analysis does not apply. For example, if the game is played with the numbers on ten one-dollar bills, and your first draw is a bill with a number that begins with 9, your best strategy is to keep the bill. For similar reasons, the strategy in playing Googol is not strictly applicable to the unmarried girl problem, as many readers pointed out, because the girl presumably has a fair knowledge of the range in value of her suitors, and has certain standards in mind. If the first man who proposes comes very close to her ideal, wrote Joseph P. Robinson, "she would have rocks in her head if she did not accept at once."

Fox and Marnie apparently hit independently on a problem that had occurred to others a few years before. A number of readers said they had heard the problem before 1958 — one recalled working on it in 1955 — but I was unable to find any published reference to it. The problem of maximizing the *value* of the selected object (rather than the chance of getting the object of highest value) seems first to have been proposed by the famous mathematician Arthur Cayley in 1875. (See Leo Moser, "On a Problem of Cayley," in *Scripta Mathematica*, September-December 1956, pages 289–292.)

4. Let 1 be the length of the square of cadets and also the time it takes them to march this length. Their speed will also be 1. Let x be the total distance traveled by the dog and also its speed. On the dog's forward trip his speed relative to the cadets will be $x - 1$. On the return trip his speed relative to the cadets will be $x + 1$. Each trip is a distance of 1 (relative to the cadets), and the two trips are completed in unit time, so the following equation can be written:

$$\frac{1}{x-1} + \frac{1}{x+1} = 1$$

This can be expressed as the quadratic: $x^2 - 2x - 1 = 0$, for which x has the positive value of $1 + \sqrt{2}$. Multiply this by 50 to get the final answer: 120.7+ feet. In other words, the dog travels a total distance equal to the length of the square of cadets plus that same length times the square root of 2.

Loyd's version of the problem, in which the dog trots *around* the moving square, can be approached in exactly the same way. I paraphrase a clear, brief solution sent by Robert F. Jackson of the Computing Center at the University of Delaware.

As before, let 1 be the side of the square and also the time it takes the cadets to go 50 feet. Their speed will then also be 1. Let x be the distance traveled by the dog and also his speed. The dog's speed with respect to the speed of the square will be $x - 1$ on his forward trip, $\sqrt{x^2 - 1}$ on each of his two transverse trips, and $x + 1$ on his backward trip. The circuit is completed in unit time, so we can write this equation:

$$\frac{1}{x-1} + \frac{2}{\sqrt{x^2-1}} + \frac{1}{x+1} = 1$$

This can be expressed as the quadratic equation: $x^4 - 4x^3 - 2x^2 + 4x + 5 = 0$. Only one positive real root is not extraneous: 4.18112+. We multiply this by 50 to get the desired answer: 209.056+feet.

Theodore W. Gibson, of the University of Virginia, found that the first form of the above equation can be written as follows, simply by taking the square root of each side:

$$\frac{1}{\sqrt{x-1}} + \frac{1}{\sqrt{x+1}} = 1$$

which is remarkably similar to the equation for the first version of the problem.

Many readers sent analyses of variations of this problem: a square formation marching in a direction parallel to the square's diagonal, formations of regular polygons with more than four sides, circular formations, rotating formations, and so on. Thomas J. Meehan and David Salsburg each pointed out that the problem is the same as that of a destroyer making a square search pattern around a moving ship, and showed how easily it could be solved by vector diagrams on what the Navy calls a "maneuvering board."

5. The simplest way to fold Barr's belt so that each end is straight across and part of a rectangle of uniform thickness is shown in Figure 16. This permits the neatest roll (the seams balance the long fold) and works regardless of the belt's length or the angles at which the ends are cut.

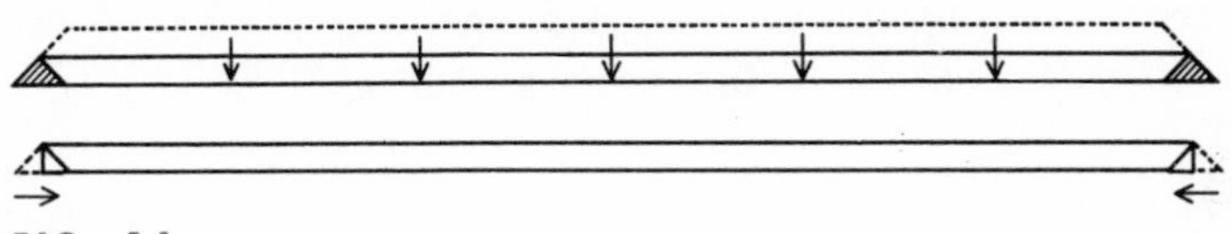

FIG. 16
How Barr folds his belt.

6. The assumption that the "lady" is Jean Brown, the stenographer, quickly leads to a contradiction. Her opening remark brings forth a reply from the person with black hair, therefore Brown's hair cannot be black. It also cannot be brown, for then it would match her name. Therefore it must be white. This leaves brown for the color of Professor Black's hair and black for Professor White. But a statement by the person with black hair prompts an exclamation from White, so they cannot be the same person.

It is necessary to assume, therefore, that Jean Brown is a man. Professor White's hair can't be white (for then it would match his or her name), nor can it be black because he (or she) replies to the black-haired person. Therefore it must be brown. If the lady's hair isn't brown, then Professor White is not a lady. Brown is a man, so Professor Black must be the lady. Her hair can't be black or brown, so she must be a platinum blonde.

7. Since the wind boosts the plane's speed from A to B and retards it from B to A, one is tempted to suppose that these forces balance each other so that total travel time for the combined flights will remain the same. This is not the case, because the time during which the plane's speed is boosted is shorter than the time during which it is retarded, so the over-all effect is one of retardation. The total travel time in a wind of constant speed and direction, regardless of the speed or direction, is always greater than if there were no wind.

8. Let x be the number of hamsters originally purchased and also the number of parakeets. Let y be the number of hamsters among the seven unsold pets. The number of parakeets among the seven will then be $7 - y$. The number of hamsters sold (at a price of \$2.20 each, which is a markup of 10 per cent over cost) will be $x - y$, and the number of parakeets sold (at \$1.10 each) will be $x - 7 + y$.

The cost of the pets is therefore $2x$ dollars for the hamsters and x dollars for the parakeets — a total of $3x$ dollars. The hamsters that were sold brought $2.2\ (x - y)$ dollars and the parakeets sold brought $1.1\ (x - 7 + y)$ dollars — a total of $3.3x - 1.1y - 7.7$ dollars.

We are told that these two totals are equal, so we equate them and simplify to obtain the following Diophantine equation with two integral unknowns:

$$3x = 11y + 77$$

Since x and y are positive integers and y is not more than 7, it is a simple matter to try each of the eight possible values (including zero) for y to determine which of them makes x also integral. There are only two such values: 5 and 2. Each would lead to a solution of the problem were it not for the fact that the parakeets were bought in pairs. This eliminates 2 as a value for y because it would give x (the number of parakeets purchased) the odd value of 33. We conclude therefore that y is 5.

A complete picture can now be drawn. The shop owner bought 44 hamsters and 22 pairs of parakeets, paying altogether \$132 for them. He sold 39 hamsters and 21 pairs of parakeets for a total of \$132. There remained five hamsters worth \$11 retail and two parakeets worth \$2.20 retail — a combined value of \$13.20, which is the answer to the problem.

CHAPTER FOUR

□

The Games and Puzzles of Lewis Carroll

THE REVEREND Charles L. Dodgson, who wrote immortal fantasy under the pseudonym of Lewis Carroll, was an undistinguished mathematician who delivered dull lectures at Oxford and penned equally dull treatises on such topics as geometry and algebraic determinants. Only when he approached mathematics in a less serious mood did his subject and his way of writing about it take on lasting interest. Bertrand Russell has said that Carroll's only significant discoveries were two logical paradoxes that were published as jokes in the journal *Mind.* Carroll also wrote two books on logic for young people, each dealing with what are now old-fashioned topics, but containing exercise problems so quaint and absurd that both books, recently combined into one Dover paperback, are now winning new readers. His serious textbooks have long been out of print, but his two volumes of original puzzles, *A Tangled Tale* and *Pillow Problems,* are also available today in a single Dover paperback edition. Without touching on any topics in these four books, or overlapping any recreational material in Warren Weaver's fine article "Lewis Carroll: Mathematician" (*Scientific American,* April 1956), let us consider some of the Reverend Dodgson's more obscure excursions into the game and puzzle field.

In *Sylvie and Bruno Concluded,* the second part of Carroll's now almost forgotten fantasy *Sylvie and Bruno,* a German professor asks a group of house guests if they are familiar with the curious paper ring that can be formed by giving a strip a half-twist, then joining the ends:

" 'I saw one made, only yesterday,' the Earl replied. 'Muriel, my child, were you not making one, to amuse those children you had to tea?'

" 'Yes, I know that Puzzle,' said Lady Muriel. 'The Ring has only *one* surface, and only *one* edge. It's very mysterious!' "

The professor proceeds to demonstrate the close connection between the Moebius strip and another remarkable topological monstrosity, the projective plane: a one-sided surface with *no* edges. First he asks Lady Muriel for three of her handkerchiefs. Two are placed together and held up by their top corners. The top edges are sewn together, then one handkerchief is given a half-twist and the bottom edges are similarly joined. The result is of course a Moebius surface with a single edge consisting of four handkerchief edges.

The third handkerchief likewise has four edges that also form a closed loop. If these four edges are now sewn to the four edges of the Moebius surface, the professor explains, the result will be a closed, edgeless surface that is like that of a sphere except that it will have only one side.

" '*I* see!' Lady Muriel eagerly interrupted. 'Its *outer* surface will be continuous with its *inner* surface! But it will take time. I'll sew it up after tea.' She laid aside the bag, and resumed her cup of tea. 'But why do you call it Fortunatus's Purse, Mein Herr?'

"The dear old man beamed upon her. . . . 'Don't you see, my child. . . . Whatever is *inside* that Purse, is *outside* it; and whatever is *outside* it, is *inside* it. So you have all the wealth of the world in that leetle Purse!' "

It is just as well that Lady Muriel never gets around to sewing on the third handkerchief. It cannot be done without self-intersection of the surface, but the proposed construction does give a valuable insight into the structure of the projective plane.

Admirers of Count Alfred Korzybski, who founded general semantics, are fond of saying that "the map is not the territory." Carroll's German professor explains how in his country a map

Lewis Carroll: a drawing by Harry Furniss, illustrator of Carroll's Sylvie and Bruno.

and territory eventually became identical. To increase accuracy, map makers gradually expanded the scale of their maps, first to six yards to the mile, then 100 yards.

" 'And then came the grandest idea of all! We actually made a map of the country, on the scale of *a mile to the mile*!'

" 'Have you used it much?' I enquired.

" 'It has never been spread out, yet,' said Mein Herr. 'The farmers objected: they said it would cover the whole country, and shut out the sunlight! So we now use the country itself, as its own map, and I assure you it does nearly as well.' "

All this is Carroll's way of poking fun at what he thought was an excessive English respect for German erudition. "Nowadays," he wrote elsewhere, "no man of Science, that setteth any store by his good name, will cough otherwise than thus, Ach! Euch! Auch!"

In *The Diaries of Lewis Carroll,* published by the Oxford University Press in 1954, are many entries that reflect his constant preoccupation with recreational mathematics. On December 19, 1898, he wrote: "Sat up last night till 4 A.M., over a tempting problem, sent me from New York, 'to find three equal [in area] rational-sided right-angled triangles.' I found *two,* whose sides are 20, 21, 29; 12, 35, 37; but could not find three." Perhaps some readers will enjoy seeing if they can succeed where Carroll failed. Actually there is no limit to the number of right triangles that can be found with integral sides and equal areas, but beyond three triangles the areas are never less than six-digit numbers. Carroll came very close to finding three such triangles, as we will explain in the answer section. There is one answer in which the area involved, although greater than the area of each triangle cited by Carroll, is still less than 1,000.

"I have worked out in the last few days," Carroll records on May 27, 1894, "some curious problems on the plan of 'lying' dilemma. E.g., 'A says B lies; B says C lies; C says A and B lie.' " The question is: Who lies and who tells the truth? One must assume that A refers to B's statement, B to C's statement, and C to the combined statements of A and B.

Of several unusual word games invented by Carroll, the solitaire game of Doublets became the most popular in his day, partly because of prize competitions sponsored by the English magazine *Vanity Fair.* The idea is to take two appropriate words of the same length, then change one to the other by a series of intermediate words, each differing by only one letter from the word preceding. Proper names must not be used for the linking words, and the words should be common enough to be found in the average abridged dictionary. For example, PIG can be turned into STY as follows:

PIG
WIG
WAG
WAY
SAY
STY

One must strive, of course, to effect the change with the smallest possible number of links. For readers who enjoy word puzzles, here are six Doublets from *Vanity Fair*'s first contest. It will be interesting to see if any readers succeed in making the changes with fewer links. The Doublets are:

Prove GRASS to be GREEN.
Evolve MAN from APE.
Raise ONE to TWO.
Change BLUE to PINK.
Make WINTER SUMMER.
Put ROUGE on CHEEK.

Like so many mathematicians, Carroll enjoyed all sorts of wordplay: composing anagrams on the names of famous people (one of his best: William Ewart Gladstone — Wild agitator! Means well), writing acrostic verses on the names of little girls, inventing riddles and charades, making puns. His letters to his child friends were filled with this sort of thing. In one letter he mentions his discovery that the letters ABCDEFGI can be rearranged to make a hyphenated word. Can anyone discover it?

Carroll's writings abound in puns, though they incline to be clever rather than outrageous. He once defined a "sillygism" as the combining of two prim misses to yield a delusion. His virtuosity in mathematical punning reached its highest point in a pamphlet of political satire entitled *Dynamics of a Parti-cle.* It opens with the following definitions:

"Plain Superficiality is the character of a speech, in which any two points being taken, the speaker is found to lie wholly with regard to those two points. Plain Anger is the inclination of two voters to one another, who meet together, but whose views are not in the same direction. When a Proctor, meeting another Proctor, makes the votes on one side equal to those on the other, the feeling entertained by each side is called Right Anger. When two parties, coming together, feel a Right Anger, each is *said* to be Complementary to the other (though, strictly speaking, this is very seldom the case). Obtuse Anger is that which is greater than Right Anger."

Mathematical puns also provide most of the humor for another Carroll pamphlet, *The New Method of Evaluation as Applied to* π. Pi stands for the salary of Benjamin Jowett, professor of Greek and translator of Plato, whom many suspected of harboring unorthodox religious views. The tract satirizes the failure of Oxford officials to agree on Professor Jowett's salary. The following passage, in which J stands for Jowett, will convey the pamphlet's flavor:

"It had long been perceived that the chief obstacle to the evaluation of π was the presence of J, and in an earlier age of mathematics J would probably have been referred to rectangular axes, and divided into two unequal parts — a process of arbitrary elimination which is now considered not strictly legitimate."

One can almost hear the Queen of Hearts screaming: "Off with his head!"

Great writers who like to indulge in wordplay are almost always admirers of Carroll. There are many Carrollian references in James Joyce's *Finnegans Wake,* including one slightly blasphemous reference to Carroll himself: "Dodgfather, Dodgson & Coo." It is not surprising to learn that Vladimir Nabokov, whose novel *Lolita* is notable not only for its startling theme but also for its verbal high jinks, translated *Alice's Adventures in Wonderland* into Russian in 1923 (not the first translation, but the best, he has said). There are other interesting Carroll-Nabokov links. Like Carroll, Nabokov is fond of chess (one of his novels, *The Defense,* is about a monomaniacal chess player) and Humbert Humbert, the narrator of *Lolita,* resembles Carroll in his enthusiasm for little girls. One must hasten to add that Carroll would surely have been shocked by *Lolita.*

Dodgson considered himself a happy man, but there is a gentle undertow of sadness that runs beneath much of his nonsense: the loneliness of a shy, inhibited bachelor who lay awake at night battling what he called "unholy thoughts" by inventing complicated "pillow problems" and solving them in his head.

Yet what are all such gaieties to me
Whose thoughts are full of indices and surds?
$x^2 + 7x + 53$
$= 11/3.$

ADDENDUM

LEWIS CARROLL invented Doublets at Christmas 1877 for two girls who "had nothing to do." He published a number of leaflets and pamphlets about the game, which he first called Word-links. For details on these publications, and a history of the game, see *The Lewis Carroll Handbook,* edited by Roger L. Green, revised edition, Oxford Press, pages 94–101.

Doublet problems appear in scores of old and new puzzle books. Dmitri Borgmann, on page 155 of his recent *Language on Vacation* (Scribner's, 1965), calls them "word ladders" and points out that the ideal word ladder is one in which the two words have no identical letters at the same positions, and the change is accomplished with the same number of steps as there are letters in each word. He gives as an example COLD to WARM in four steps.

It is not surprising to find Doublets turning up (under the name of "word golf") in Nabokov's *Pale Fire.* The novel's mad narrator, commenting on line 819 of the poem around which the novel is woven, speaks of HATE to LOVE in three steps, LASS to MALE in four, and LIVE to DEAD in five, with LEND in the middle. Solutions to the first two are provided by Mary McCarthy in her remarkable review of the novel (*New Republic,* June 4, 1962). Miss McCarthy adds some new Doublets of her own, based on the words in the novel's title.

John Maynard Smith, in an essay on "The Limitations of Molecular Evolution" (in *The Scientist Speculates,* edited by I. J. Good, Basic Books, 1962, pages 252–256), finds a striking resemblance between Doublets and the process by which one species evolves into another. If we think of the helical DNA molecule as one enormously long "word," then single mutations correspond to steps in the word game. APE actually changes to MAN by a process closely analogous to the playing of Doublets! Smith gives as an example the ideal change of WORD to GENE in four steps.

ANSWERS

THE ANSWER in smallest numbers for Lewis Carroll's problem of finding three right triangles with integral sides and equal areas is 40, 42 and 58; 24, 70 and 74; and 15, 112 and 113. In each case the area is 840. Had Carroll doubled the size of the two triangles that he found, he would have obtained the first two triangles cited above, from which the step to the third would have been easy. Henry Ernest Dudeney, in the answer to problem 107 in his *Canterbury Puzzles*, gives a formula by which such triangle triplets can be easily found.

Carroll's truth-and-lie problem has only one answer that does not lead to a logical contradiction: A and C lie; B speaks the truth. The problem yields easily to the propositional calculus by taking the word "says" as the logical connective called equivalence. Without drawing on symbolic logic one can simply list the eight possible combinations of lying and truth-telling for the three men, then explore each combination, eliminating those that lead to logical contradictions.

Carroll's solutions to the six Doublets are: GRASS, CRASS, CRESS, TRESS, TREES, FREES, FREED, GREED, GREEN; APE, ARE, ERE, ERR, EAR, MAR, MAN; ONE, OWE, EWE, EYE, DYE, DOE, TOE, TOO, TWO; BLUE, GLUE, GLUT, GOUT, POUT, PORT, PART, PANT, PINT, PINK; WINTER, WINNER, WANNER, WANDER, WARDER, HARDER, HARPER, HAMPER, DAMPER, DAMPED, DAMMED, DIMMED, DIMMER, SIMMER, SUMMER; ROUGE, ROUGH, SOUGH, SOUTH, SOOTH, BOOTH, BOOTS, BOATS, BRATS, BRASS, CRASS, CRESS, CREST, CHEST, CHEAT, CHEAP, CHEEP, CHEEK.

The letters ABCDEFGI rearrange to make the hyphenated word BIG-FACED.

After Carroll's answers to his Doublets appeared in *Scientific American*, a large number of readers sent in shorter answers. The following beautiful seven-step change of GRASS to GREEN was discovered by A. L. Cohen, Scott Gallagher, Lawrence Jaseph, George Kapp, Arthur H. Lord, Sidney J. Osborn and H. S. Percival:

GRASS
CRASS
CRESS
TRESS
TREES
TREED
GREED
GREEN

Mrs. C. C. Gotlieb sent a similar seven-stepper in which the second, third and fourth words of the above solution are replaced by GRAYS, TRAYS, and TREYS. If the archaic word GREES is accepted, the change can be made in four steps, as Stephen Barr, H. S. Percival and Richard D. Thurston independently found:

GRASS
GRAYS
GREYS
GREES
GREEN

Ten readers (David M. Bancroft, Robert Bauman, Frederick J. Hooven, Arthur H. Lord, Mrs. Henry A. Morss, Sidney J. Osborn, Dodi Schultz, George Starbuck, Edward Wellen and a reader whose signature was illegible) sent the following excellent five-step change of APE to MAN:

APE
APT
OPT
OAT
MAT
MAN

Many readers found seven-step changes of ONE to TWO, but since all contained at least one uncommon word, I award the palm to H. S. Percival for this six-stepper:

ONE
OYE
DYE
DOE
TOE
TOO
TWO

"Oye" is a Scottish word for grandchild, but it appears in *Webster's New Collegiate Dictionary.*

BLUE was turned to PINK in seven steps by Wendell Perkins (left) and Richard D. Thurston (right):

BLUE	BLUE
GLUE	BLAE
GLUT	BLAT
GOUT	BEAT
POUT	PEAT
PONT	PENT
PINT	PINT
PINK	PINK

Frederick J. Hooven found this admirable eight-step change, all with common words, of WINTER to SUMMER:

WINTER
WINDER
WANDER
WARDER
HARDER
HARMER
HAMMER
HUMMER
SUMMER

But it can be done in seven steps by using less familiar words (Mrs. Henry A. Morss, Richard D. Thurston, and H. S. Percival):

WINTER
LINTER
LISTER
LISPER
LIMPER
SIMPER (or LIMMER)
SIMMER
SUMMER

Lawrence Jaseph (left) and Frederick J. Hooven (right) reduced the change of ROUGE to CHEEK to 11 steps:

ROUGE	ROUGE
ROUTE	ROUTE
ROUTS	ROUTS
ROOTS	ROOTS
BOOTS	COOTS
BLOTS	COONS
BLOCS	COINS
BLOCK	CHINS
CLOCK	CHINK
CHOCK	CHICK
CHECK	CHECK
CHEEK	CHEEK

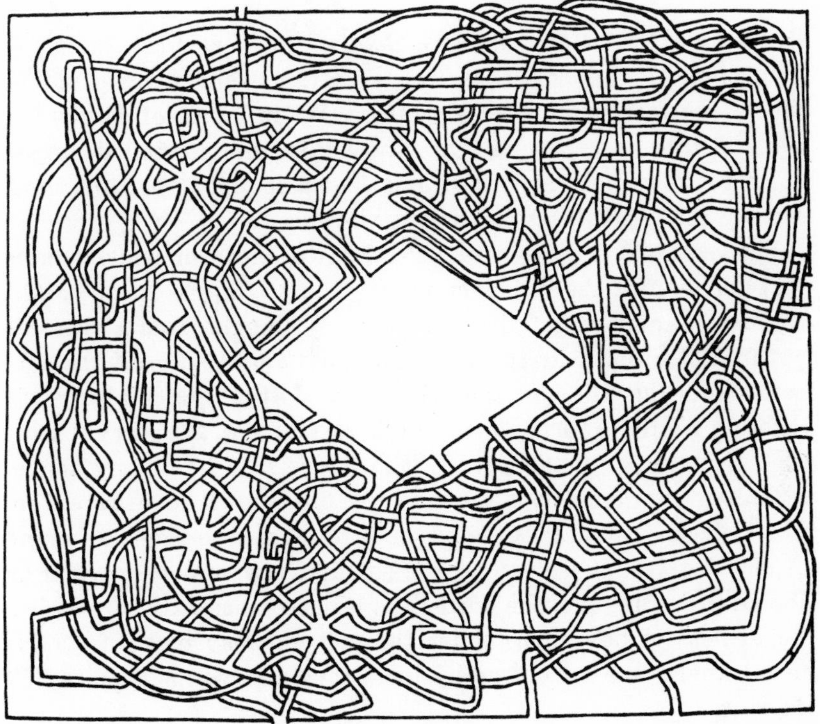

A maze drawn by Lewis Carroll in his early twenties. The object is to find your way out of the central space. Paths cross over and under one another, but are occasionally blocked by single-line barriers.

CHAPTER FIVE

□

Paper Cutting

IN *The 2nd Scientific American Book of Mathematical Puzzles & Diversions* there is a chapter on recreations that involve folding sheets of paper without cutting them. When a pair of scissors is brought into play, a wealth of interesting new possibilities open up, many of which serve to dramatize basic and important theorems of plane geometry in curious ways.

For example, consider the well-known theorem which states that the sum of the interior angles of any triangle is a straight angle (an angle of 180 degrees). Cut a triangle from a sheet of paper. Put a dot near the vertex of each angle, snip off the corners, and you will find that the three dotted angles always fit together neatly to form a straight angle [*see Fig. 17a*]. Try it with the corners of a quadrilateral. The figure may be of any shape, including concave forms such as the one shown in Figure 17b. The four snipped angles always join to form a perigon: an angle of 360 degrees. If we extend the sides of any convex polygon as shown in Figure 17c, the dotted angles are called exterior angles. Regardless of how many sides the polygon may have, if its exterior angles are cut out and joined, they also will add up to 360 degrees.

If two or more sides of a polygon intersect, we have what is sometimes called a crossed polygon. The five-pointed star or pentagram, the fraternal symbol of the ancient Pythagoreans, is

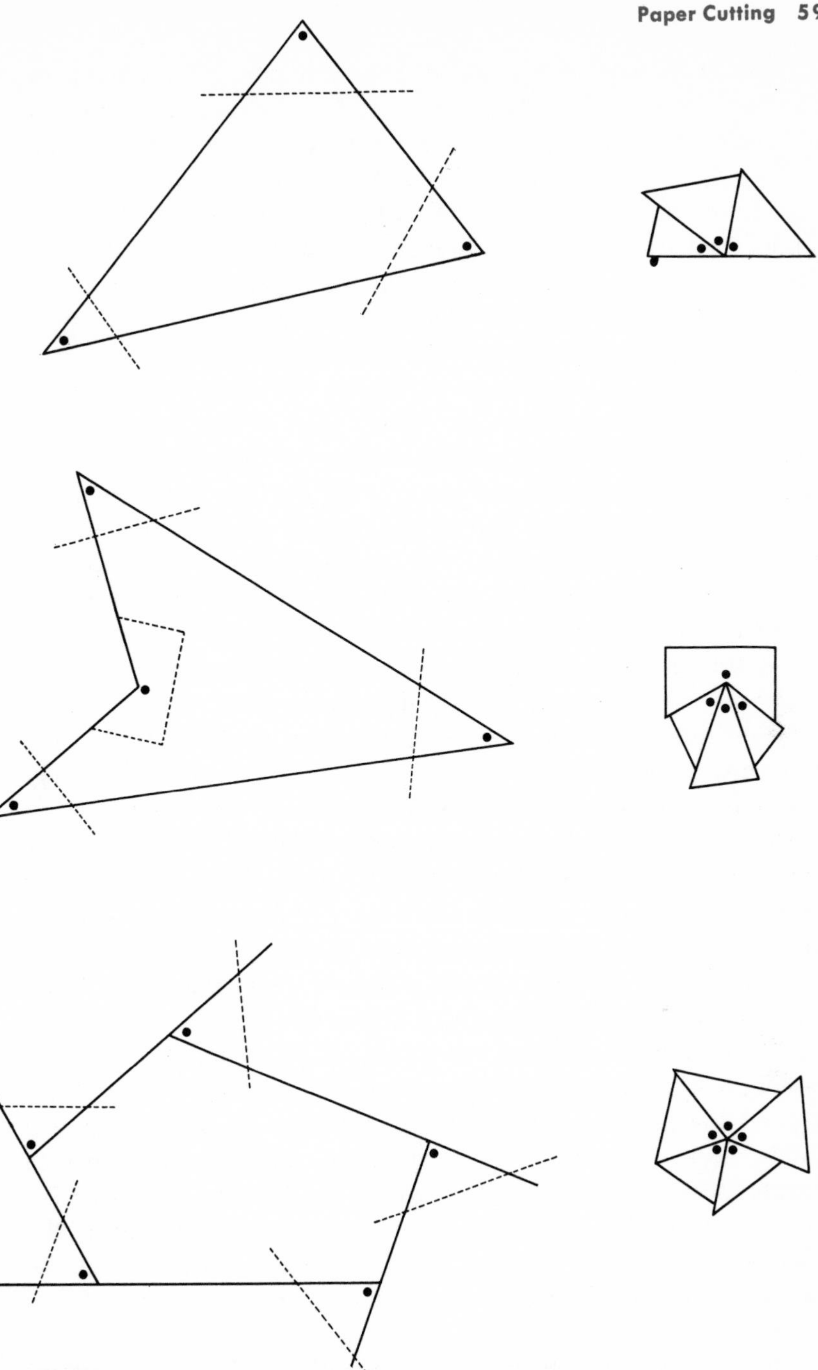

FIG. 17
How to discover theorems of plane geometry by cutting polygons.

a familiar example. Rule the star as irregularly as you please (you may even include the degenerate forms shown in Figure 18, in which one or two points of the star fail to extend beyond the body), dot the five corners, cut out the star and trim off the corners. You may be surprised to find that, as in the case of the triangle, the points of any pentagram join to form a straight angle. This theorem can be confirmed by another quaint empirical technique that might be called the sliding-match method. Draw a large pentagram, then place a match alongside one of the lines as shown in the top illustration of Figure 18. Slide the match up until its head touches the top vertex, then swing its tail to the left until the match is alongside the other line. The match has now altered its orientation on the plane by an angle equal to the angle at the top corner of the star. Slide the match down to the next corner and do the same thing. Continue sliding the match around the star, repeating this procedure at each vertex. When the match is back to its original position, it will be upside down, having made a clockwise rotation of exactly 180 degrees. This rotation is clearly the sum of the pentagram's five angles.

The sliding-match method can be used for confirming all of the theorems mentioned, as well as for finding new ones. It is a handy device for measuring the angles of any type of polygon, including the star forms and the helter-skelter crossed varieties. Since the match must return to its starting position either pointing the same way or in the opposite direction, it follows (providing the match has always rotated in the same direction) that the sum of the traversed angles must be a multiple of a straight angle. If the match rotates in both directions during its trip, as is often the case with crossed polygons, we cannot obtain a sum of the angles, although other theorems can be stated. For instance, a match slid around the perimeter of the crossed octagon in Figure 19 will rotate clockwise at the angles marked A, and the same distance counterclockwise at the angles marked B. Thus we cannot arrive at the sum of the eight angles, but we can say that the sum of the four A angles equals the sum of the four B angles. This can be easily verified by the scissors method or by a formal geometrical proof.

The familiar Pythagorean theorem lends itself to many elegant scissors-and-paper demonstrations. Here is a remarkable one dis-

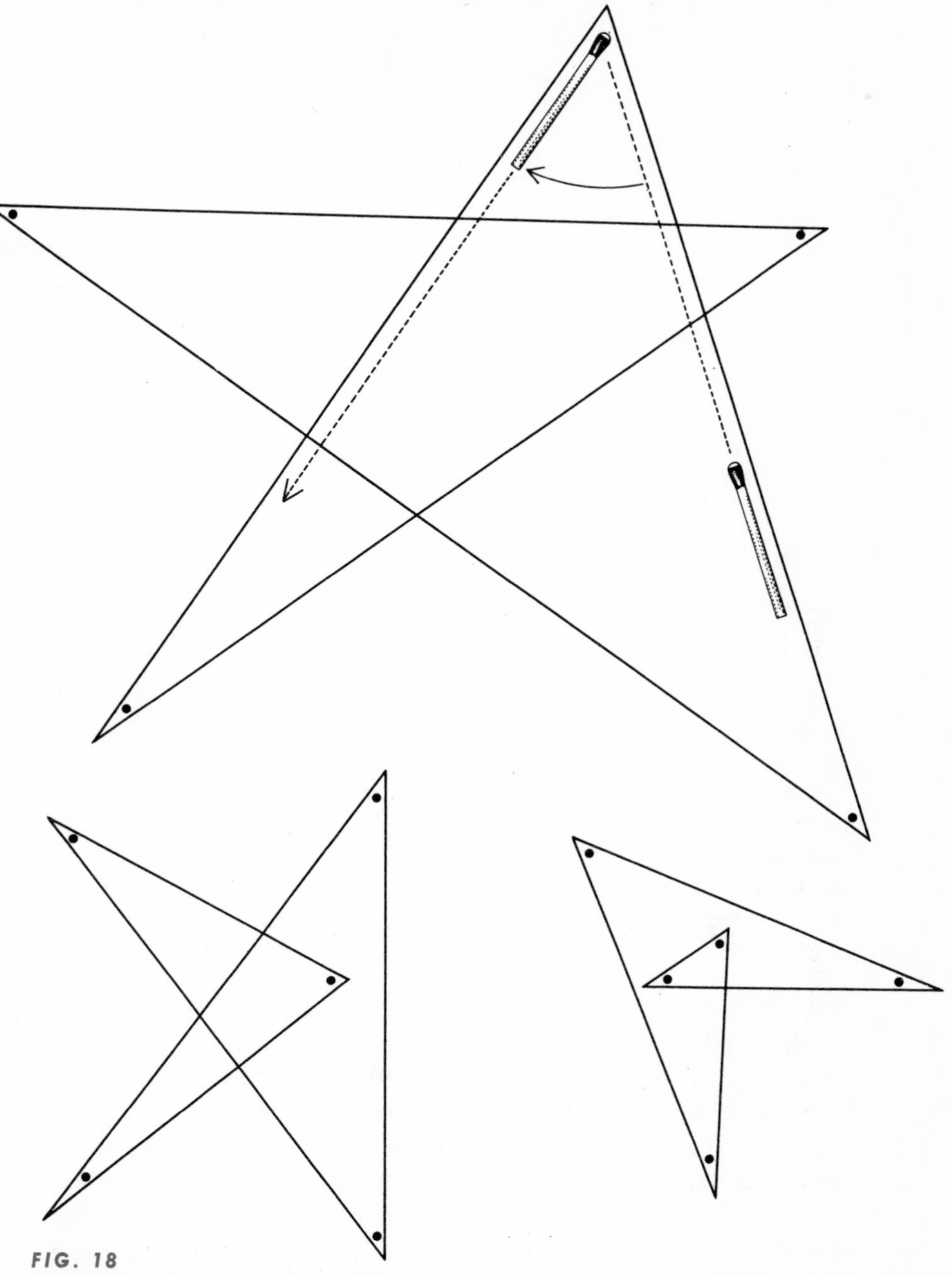

FIG. 18
Sliding a match around pentagrams shows that the dotted angles add up to 180 degrees.

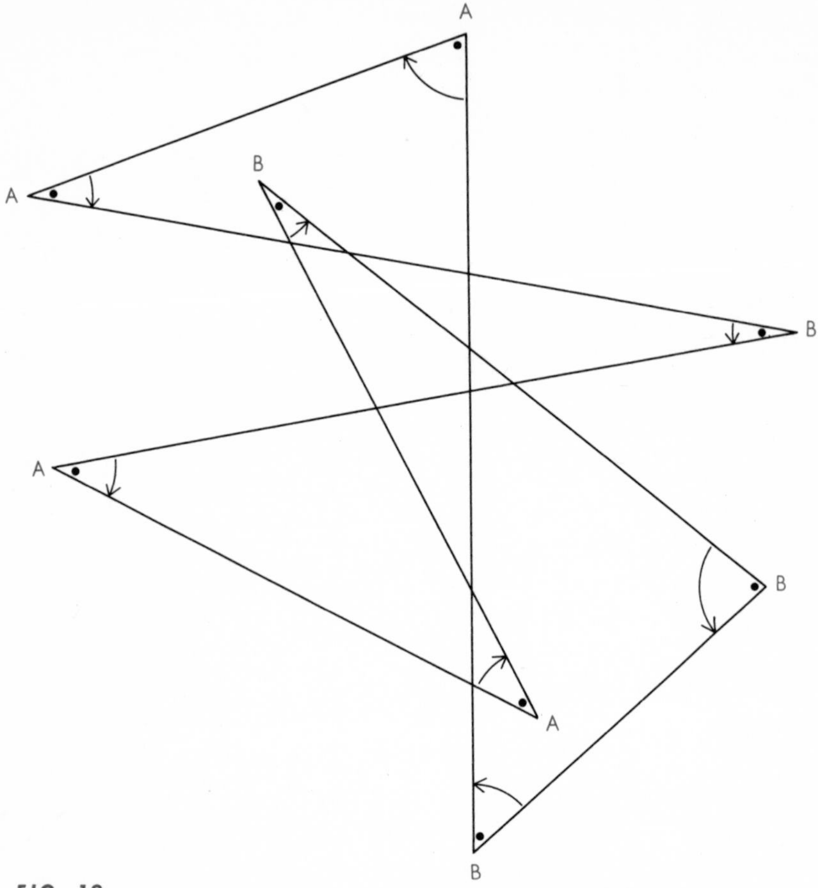

FIG. 19
On this crossed octagon the sum of the angles marked A equals the sum of those marked B.

covered in the 19th century by Henry Perigal, a London stockbroker and amateur astronomer. Construct squares on the two legs of any right triangle [*see Fig. 20*]. Divide the larger square (or either square if they are the same size) into four identical parts by ruling two lines through the center, at right angles to each other and with one line parallel to the triangle's hypotenuse. Cut out the four parts and the smaller square. You will find that all five pieces can be shifted in position, without changing their orientation on the plane, to form one large square (shown by broken lines) on the hypotenuse.

Perigal discovered this dissection in about 1830, but did not publish it until 1873. He was so delighted with it that he had the diagram printed on his business card, and gave away hundreds of puzzles consisting of the five pieces. (Someone who has not seen the diagram will have considerable difficulty fitting the pieces together, first to make two squares, then one large square.) It is amusing to learn from Perigal's obituary, in the 1899 notices of the Royal Astronomical Society of London, that his "main astronomical aim in life" was to convince others, "especially young men not hardened in the opposite belief," that it was a

FIG. 20
Henry Perigal's scissors-and-paper demonstration of Euclid's famous 47th proposition.

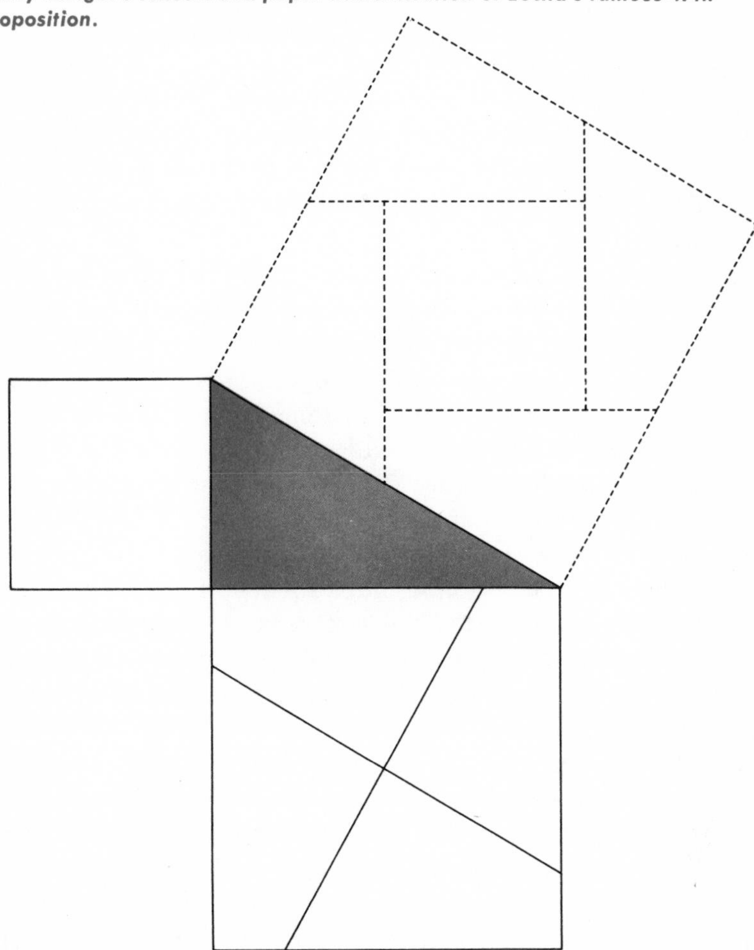

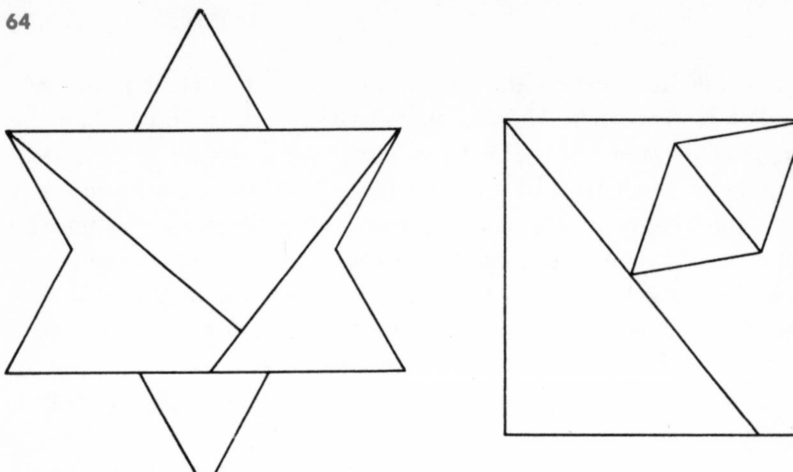

FIG. 21
E. B. Escott discovered this dissection of a regular hexagram to a square.

grave misuse of words to say that the moon "rotates" as it revolves around the earth. He wrote pamphlets, built models and even composed poems to prove his point, "bearing with heroic cheerfulness the continual disappointment of finding none of them of any avail."

The dissection of polygons into pieces that form other polygons is one of the most fascinating branches of recreational mathematics. It has been proved that any polygon can be cut into a finite number of pieces that will form any other polygon of the same area, but of course such dissections have little interest unless the number of pieces is small enough to make the change startling. Who would imagine, for example, that the regular hexagram, or six-pointed Star of David, could be cut [*see Fig. 21*] into as few as five pieces that will form a square? (The regular pentagram cannot be dissected into a square with less than eight pieces.) Harry Lindgren, of the Australian patent office, is perhaps the world's leading expert on dissections of this type. In Figure 22 we see his beautiful six-piece dissection of a regular dodecagon to a square.

A quite different class of paper-cutting recreation, more familiar to magicians than mathematicians, involves folding a sheet of paper several times, giving it a single straight cut, then opening up one or both of the folded pieces to reveal some sort of surprising result. For example, the unfolded piece may prove to be a regular geometric figure or design, or it may have a hole

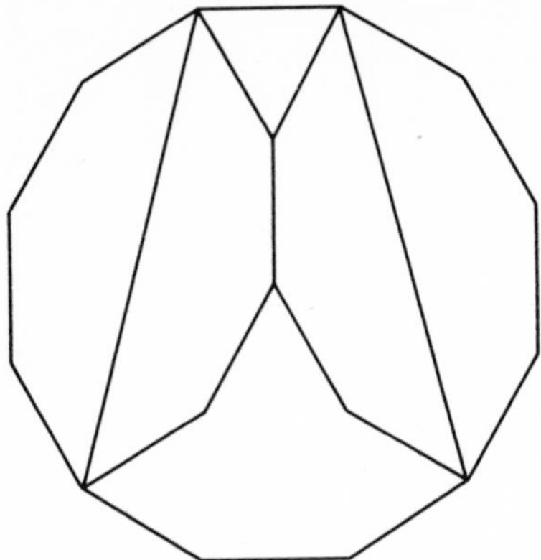
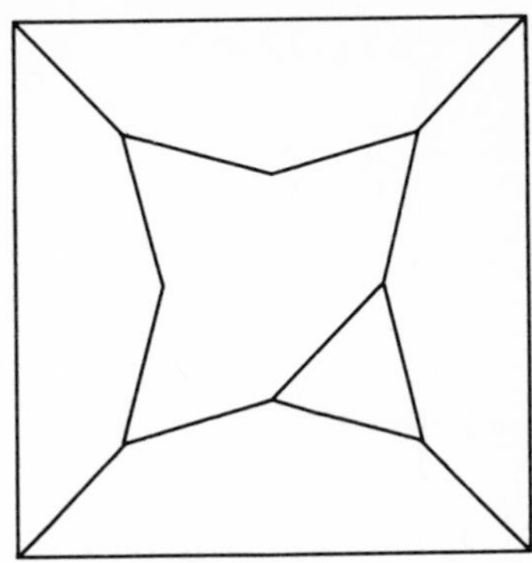

FIG. 22
Harry Lindgren's dissection of a regular dodecagon to a square.

with such a shape. In 1955 the Ireland Magic Company of Chicago published a small book called *Paper Capers,* by Gerald M. Loe, which deals almost entirely with such stunts. The book explains how to fold a sheet so that a single cut will produce any desired letter of the alphabet, various types of stars and crosses, and such complex patterns as a circular chain of stars, a star within a star, and so on. An unusual single-cut trick that is familiar to American magicians is known as the bicolor cut. A square of tissue paper, colored red and black to look like an eight-by-eight checkerboard, is folded a certain way, then given a single straight snip. The cut separates the red squares from the black and simultaneously cuts out each individual square. With a sheet of onionskin paper (the thin paper makes it possible to see outlines through several thicknesses) it is not difficult to devise a method for this trick, as well as methods for single-cutting simple geometrical figures; but more complicated designs — the swastika for instance — present formidable problems.

An old paper-cutting stunt, of unknown origin, is illustrated in Figure 23. It is usually presented with a story about two contemporary political leaders, one admired, the other hated. Both men die and approach the gates of heaven. The Bad Guy naturally lacks the necessary sheet of paper authorizing his admittance. He seeks the aid of the Good Guy, standing just behind him. GG folds his sheet of paper as shown in *a,b,c,d* and *e,* then cuts it along the indicated dotted line. He retains the part on the right, giving

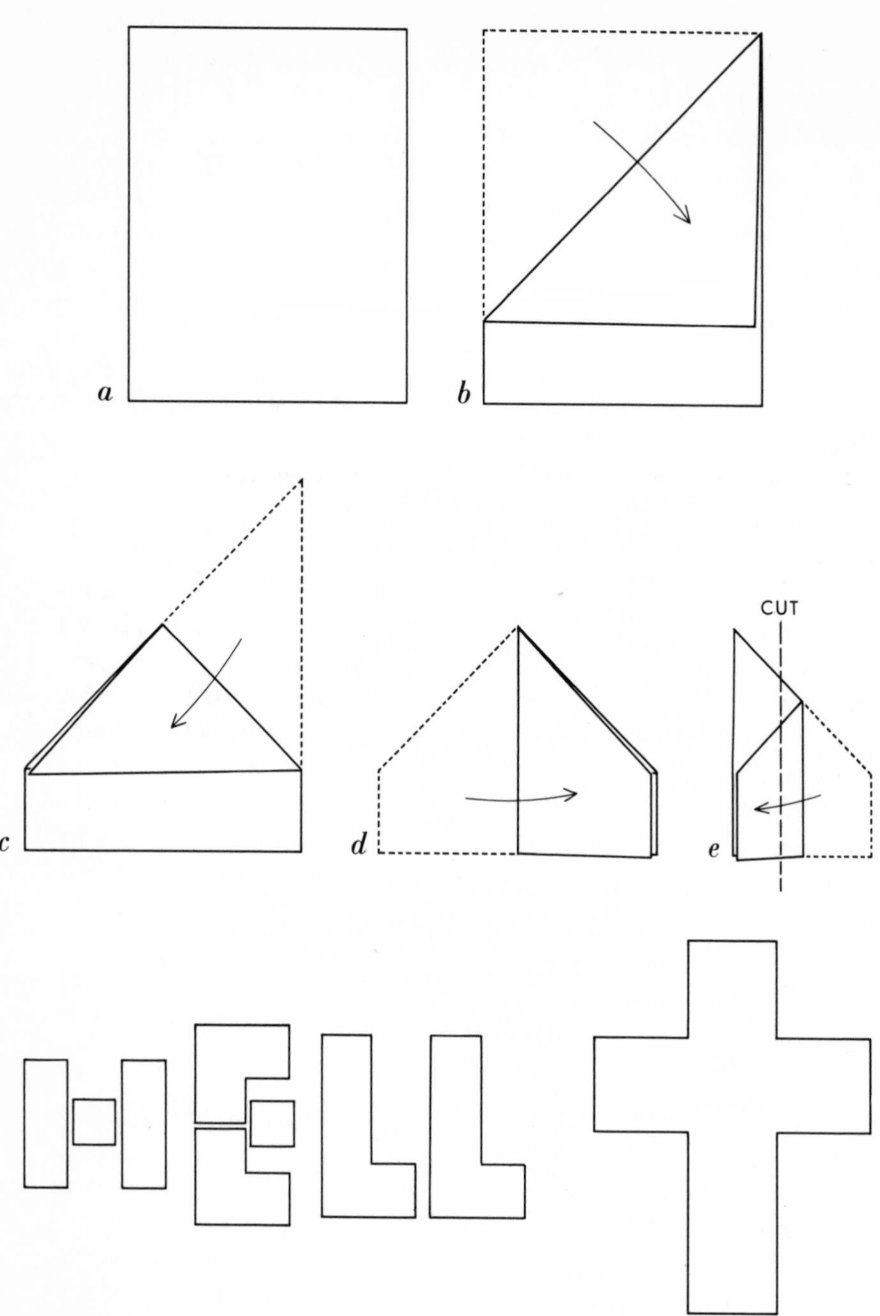

FIG. 23
An old paper-cutting trick.

the rest to BG. Saint Peter opens the BG's pieces, arranges them to form "Hell" as shown at bottom left, and sends him off. When Saint Peter opens the paper presented by the GG, he finds it in the shape of the cross shown at bottom right.

It is obviously impossible to fold a sheet flat in such a way that a straight cut will produce curved figures, but if a sheet is rolled into a cone, plane slices through it will leave edges in the form of circles, ellipses, parabolas or hyperbolas, depending on the angle of the cut. These of course are the conic sections studied by the Greeks. Less well known is the fact that a sine curve can be quickly produced by wrapping a sheet of paper many times around a cylindrical candle, then cutting diagonally through both paper and candle. When unrolled, each half of the paper will have a cut edge in the form of a sine curve, or sinusoid, one of the fundamental wave forms of physics. The trick is also useful to the housewife who wants to put a rippling edge on a sheet of shelf paper.

Here are two fascinating cut-and-fold problems, both involving cubes. The first is easy; the second, not so easy.

1. What is the shortest strip of paper one inch wide that can be folded to make all six sides of a one-inch cube?

2. A square of paper three inches wide is black on one side and white on the other. Rule the square into nine one-inch squares. By cutting only along the ruled lines, is it possible to cut a pattern that will fold along the ruled lines into a cube that is all black on the outside? The pattern must be a single piece, and no cuts or folds are permitted that are not along the lines that divide the sheet into squares.

ADDENDUM

THERE ARE, of course, all sorts of traditional geometric proofs that the points of the three different types of pentagrams shown in Figure 18 have a total of 180 degrees. The reader may enjoy working out some of them, if only to see how much simpler and more intuitively evident the sliding-match proofs are.

Perigal first published his Pythagorean dissection in *Messenger of Mathematics,* Vol. 2, new series, 1873, pages 103–106. For biographical information on Perigal, see his obituary in the

Monthly Notices of the Royal Astronomical Society of London, Vol. 59, 1899, pages 226–228. Some of his pamphlets are discussed by Augustus de Morgan in his well-known *Budget of Paradoxes* (reprinted by Dover in 1954).

The elegant hexagram-to-square dissection was discovered by Edward Brind Escott, an insurance company actuary who lived in Oak Park, Illinois, and who died in 1946. He was an expert on number theory, contributing frequently to many different mathematical journals. His hexagram dissection is given by Henry Ernest Dudeney as the solution to problem 109 in *Modern Puzzles* (1926).

For more about Lindgren's remarkable dissections, see the Mathematical Games department of *Scientific American,* November 1961, and Lindgren's book on dissections (listed in the Bibliography).

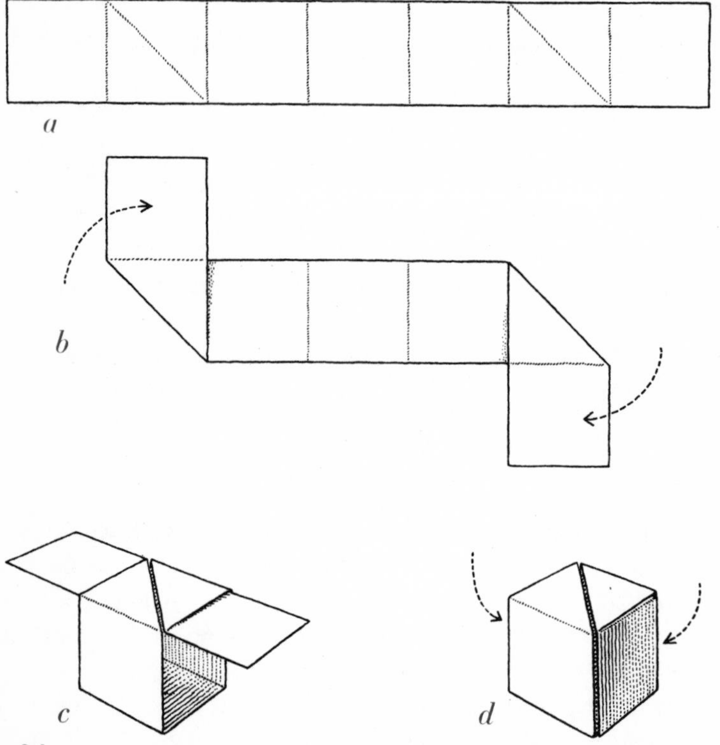

FIG. 24
How a one-inch cube can be folded from a strip one inch wide and seven inches long.

ANSWERS

THE SHORTEST strip of paper, one inch wide, that can be folded into a one-inch cube is seven inches. A method of folding is depicted in Figure 24. If the strip is black on one side, eight inches are necessary for folding an all-black cube. (A way of doing this is shown in *Recreational Mathematics Magazine*, February 1962, page 52.)

The three-inch-square sheet, black on one side only, can be cut and folded into an all-black cube in many different ways. This cannot be accomplished with a pattern of less than eight unit squares, but the missing square inch may be in any position. Figure 25 shows how a pattern with the missing square in the center can be folded into the black cube. In all solutions, the cuts have a total length of five units. (If the entire sheet is used for the pattern, the length of the cut lines can be reduced to four.)

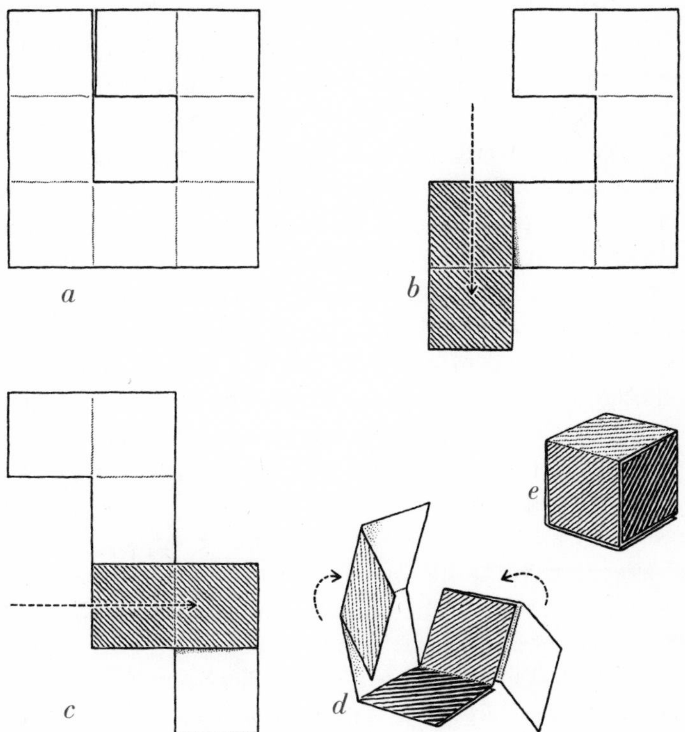

FIG. 25
An all-black cube can be folded with the pattern at top left. Pattern is black on underside.

CHAPTER SIX

□

Board Games

"GAMES POSSESS some of the qualities of works of art," Aldous Huxley has written. "With their simple and unequivocal rules, they are like so many islands of order in the vague untidy chaos of experience. When we play games, or even when we watch them being played by others, we pass from the incomprehensible universe of given reality into a neat little man-made world, where everything is clear, purposive and easy to understand. Competition adds to the intrinsic charm of games by making them exciting, while betting and crowd intoxication add, in their turn, to the thrills of competition."

Huxley is speaking of games in general, but his remarks apply with special force to mathematical board games in which the outcome is determined by pure thought, uncontaminated by physical prowess or the kind of blind luck supplied by dice, cards and other randomizing devices. Such games are as old as civilization and as varied as the wings of butterflies. Fantastic amounts of mental energy have been expended on them, considering the fact that until quite recently they had no value whatever beyond that of relaxing and refreshing the mind. Today they have suddenly become important in computer theory. Chess-playing and checker-playing machines that profit from experience may be the forerunners of electronic minds capable of developing powers as yet unimaginable.

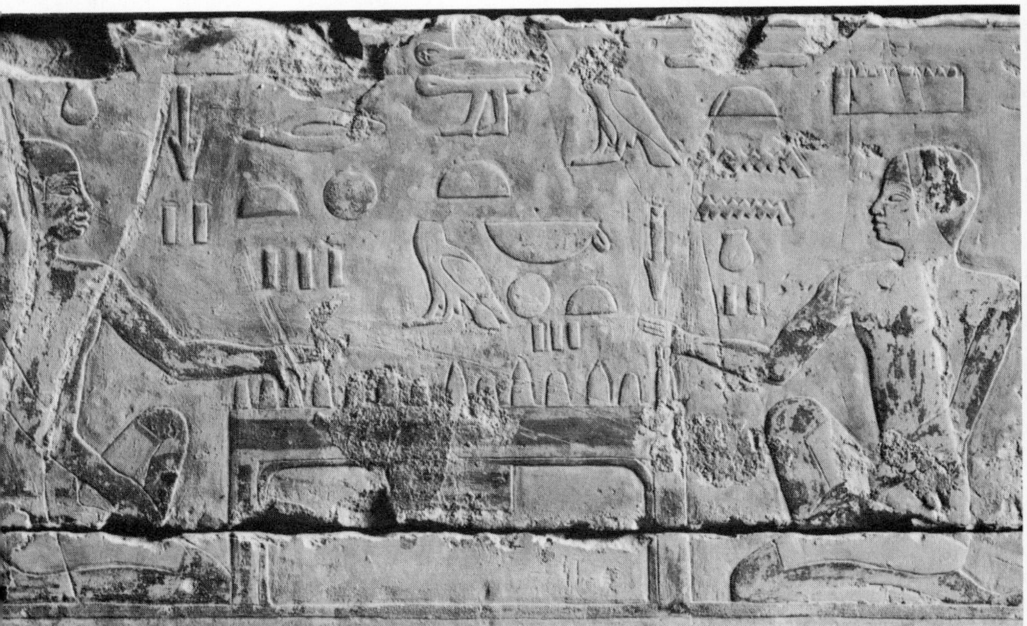

FIG. 26
Relief from a tomb at Sakkara in Egypt shows a board game in profile. Relief dates from 2500 B.C. Courtesy of The Metropolitan Museum of Art, Rogers Fund, 1908.

The earliest records of mathematical board games are found in the art of ancient Egypt, but they convey little information because of the Egyptian convention of showing scenes only in profile [*see Fig. 26*]. Some games involving boards have been found in Egyptian tombs [*Fig. 27*], but they are not board games in the strict sense, because they also involve a chance element. A bit more is known about Greek and Roman board games, but it was not until the 13th century A.D. that anyone thought it important enough to record the rules of a board game, and it was not until the 17th century that the first books on games were written.

Like biological organisms, games evolve and proliferate new species. A few simple games, such as ticktacktoe, may remain unchanged for centuries; others flourish for a time, then vanish completely. The outstanding example of a dinosaur diversion is Rithmomachy. This was an extremely complicated number game played by medieval Europeans on a double chessboard with eight

FIG. 27
Board game of senet, found in Egyptian tomb of 1400 B.C., also involved throwing sticks. The Metropolitan Museum of Art, gift of Egypt Exploration Fund, 1901.

cells on one side and sixteen cells on the other, and with pieces in the shapes of circles, squares and triangles. It traces back at least to the 12th century, and as late as the 17th century it was mentioned by Robert Burton, in *The Anatomy of Melancholy,* as a popular English game. Many learned treatises were written about it, but no one plays it today except a few mathematicians and medievalists.

In the U.S. the two most popular mathematical board games are of course checkers and chess. Both have long and fascinating histories, with unexpected mutations in rules from time to time and place to place. Today the American checkers is identical with the English "draughts," but in other countries there are wide variations. The so-called Polish checkers (actually invented in France) is now the dominant form of the game throughout most of Europe. It is played on a ten-by-ten board, each side having twenty men that capture backward as well as forward. Crowned pieces (called queens instead of kings) move like the bishop in chess, and in making a jump can land on any vacant cell beyond the captured piece. The game is widely played in France (where it is called *dames*) and in Holland, and it is the subject of a large analytical literature. In the French-speaking provinces of Canada, and in parts of India, Polish checkers is played on a twelve-by-twelve board.

German checkers (*Damenspiel*) resembles Polish checkers, but it is usually played on the English eight-by-eight board. A similar form of this "minor Polish" game, as it is sometimes called, is popular in the U.S.S.R., where it is known as *shashki.* Spanish and Italian variants also are closer to the English. Turkish checkers (*dama*) is also played on an eight-by-eight board, but each side has sixteen men that occupy the second and third rows at the outset. Pieces move and jump forward and sideways, but not diagonally, and there are other radical departures from both the English and the Polish forms.

Chess likewise has varied enormously in its rules, tracing back ultimately to an unknown origin in India, probably in the sixth century A.D. True, there is today an international chess that is standardized, but there are still many excellent non-European forms of the game that obviously share a common origin with international chess. Japanese chess (*shogi*) is played as enthusiastically in modern Japan as *go,* though only the latter game is known in Western countries. *Shogi* is played on a nine-by-nine board, with twenty men on each side, arranged at the start on the first three rows. The game is won, as in Western chess, by checkmating a piece that moves exactly like the king. An interesting feature of the game is that captured pieces can be returned to the board to be used by the captor.

Chinese chess (*tséung k'i*) also ends with the checkmate of a piece that moves like the king in Western chess, but the rules are quite different from those of the Japanese game. Its 32 pieces rest on the intersections of an eight-by-eight board that is divided across the center by a blank horizontal row called the "river." A third variant, Korean chess (*tjyang-keui*), is played on the intersections of a board that has the same pattern as the Chinese except that the "river" is not specially marked, so the board looks like an eight-by-nine checkerboard. The pieces are the same in number as the Chinese pieces, with the same names and (except for the king) the same starting positions, but the two games differ considerably in rules and the powers of the pieces. Devotees of each of the three Oriental versions of chess look upon the other two versions, as well as Western chess, as decidedly inferior.

Martian chess ("jetan"), explained by Edgar Rice Burroughs in the appendix to his novel *The Chessmen of Mars,* is an amusing variant, played on a ten-by-ten board with unusual pieces and novel rules. For example, the princess (which corresponds roughly to our king) has the privilege of one "escape move" per game that permits her to flee an unlimited distance in any direction.

In addition to these regional variants of chess, modern players, momentarily bored with the orthodox game, have invented a weird assortment of games known as fairy chess. Among the many fairy-chess games that can be played on the standard board are: two-move chess, in which each player plays twice on his turn; a game in which one side plays with no pawns, or with an extra row of pawns instead of a queen; cylindrical chess, in which the left side of the board is considered joined to the right side (if the board is thought of as having a half-twist before the sides are joined, it is called Moebius-strip chess); transportation chess, in which any piece can be moved on top of the rook and carried by the rook to another square. Dozens of strange new pieces have been introduced, such as the chancellor (combining the moves of rook and knight), the centaur (combining bishop and knight) and even neuter pieces (*e.g.*, a blue queen) that can be played by either side. (In Lewis Padgett's science-fiction novel *The Fairy Chessmen* a war is won by a mathematician who makes a hobby of fairy chess. His mind, accustomed to breaking rules, is elastic enough to cope with an equation too bizarre for his more brilliant but more orthodox colleagues.)

An amusing species of fairy chess that is quite old, but still provides a delightful interlude between more serious games, is played as follows. One player sets up his sixteen men in the usual way, but his opponent has only one piece, called the maharajah. A queen may be used for this piece, but its moves combine those of queen and knight. It is placed at the outset on any free square not threatened by a pawn; then the other side makes the first move. The maharajah loses if he is captured, and wins if he checkmates the king. Pawns are not permitted to be replaced by queens or other pieces if they advance to the last row. Without this proviso it is easy to defeat the maharajah simply by advancing the rook pawns until they can be queened. Since these and all the other pawns are protected, there is no way the maharajah can prevent both pawns from becoming queens. With three queens and two rooks in play, the game is easily won.

Even with this proviso, it might be thought that the maharajah has a poor chance of winning, but his mobility is so great that if he moves swiftly and aggressively, he often checkmates early in the game. At other times he can sweep the board clean of pieces and then force the lone king into a corner checkmate.

Hundreds of games have been invented that are played on a standard chessboard but have nothing in common with either chess or checkers. One of the best, in my opinion, is the now-forgotten game of "reversi." It uses 64 counters that have contrasting colors, say red and black, on their opposite sides. A crude set can be made by coloring one side of a sheet of cardboard, then cutting out small circles; a better set can be constructed by buying inexpensive checkers or poker chips and gluing the pieces into red-black pairs. It is worth the trouble, because the game can be an exciting one for every member of the family.

Reversi starts with an empty board. One player has 32 pieces turned red-side up; the other has 32 turned black-side up. Players alternate in placing a single man on the board in conformity with the following rules:

1. The first four men must be placed on the four central squares. Experience has shown that it is better for the first player to place his second man above, below, or to the side of his first piece (an example is shown in Figure 28), rather than diagonally adjacent, but this is not obligatory. By the same token, it is wise for the second player not to play diagonally opposite his opponent's first move, especially if his opponent is a novice. This gives the first player a chance to make the inferior diagonal move on his second play. Between experts, the game always begins with the pattern shown in Figure 28.

2. After the four central squares are filled, players continue placing single pieces. Each must be placed so that it is adjacent to a hostile piece, orthogonally or diagonally. Moreover, it must also be placed so that it is in direct line with another piece of the same color, and with one or more enemy pieces (and no vacant cells) in between. In other words, a piece must always be placed so that it is one of a pair of friendly pieces on opposite sides of

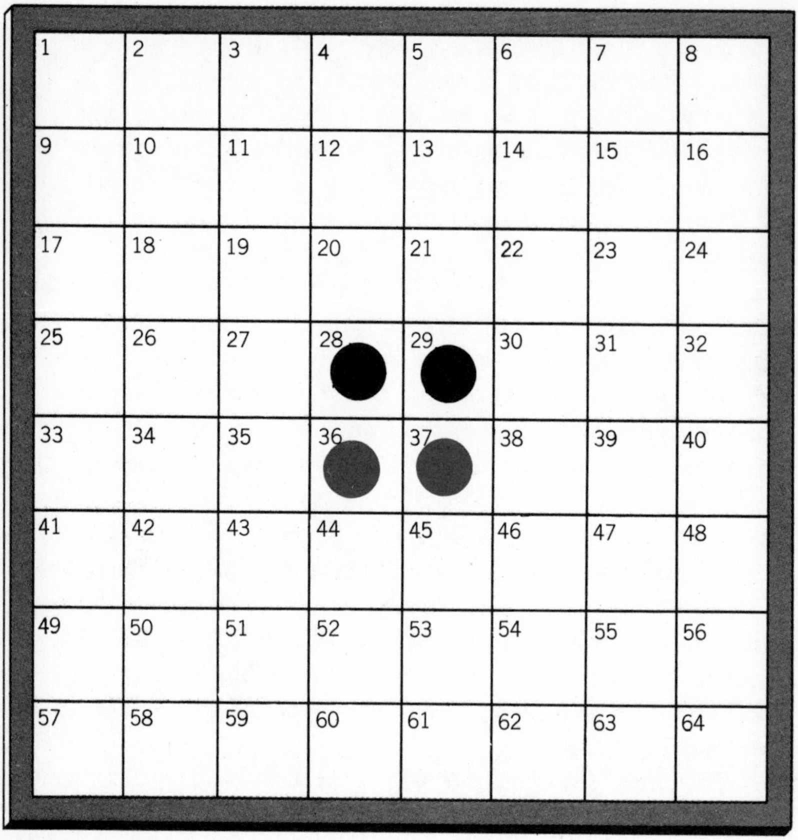

1	2	3	4	5	6	7	8
9	10	11	12	13	14	15	16
17	18	19	20	21	22	23	24
25	26	27	28	29	30	31	32
33	34	35	36	37	38	39	40
41	42	43	44	45	46	47	48
49	50	51	52	53	54	55	56
57	58	59	60	61	62	63	64

FIG. 28
An opening for the board game of reversi. Numbers are for reference only.

an enemy piece or at opposite ends of a chain of enemy pieces. The enemy pieces are considered captured, but instead of being removed they are turned over, or "reversed," so that they become friendly pieces. They are, so to speak, "brainwashed" so that they join their captors. Pieces remain fixed throughout the game, but may be reversed any number of times.

3. If the placing of a piece simultaneously captures more than one chain of enemy pieces, the pieces in both chains are reversed.

4. Pieces are captured only by the placing of a hostile piece. Chains that become flanked at both ends as a result of other causes are not captured.

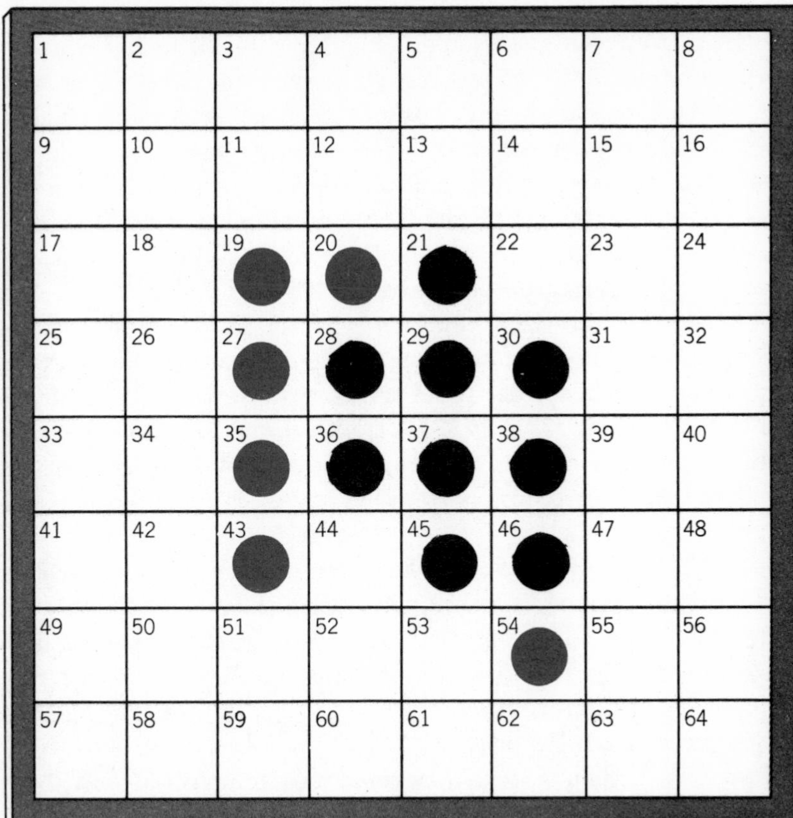

FIG. 29
If reversi player with colored pieces makes the next move, he can win six pieces.

5. If a player cannot move, he loses his turn. He continues to lose his turn until a legal move becomes possible for him.

6. The game ends when all 64 squares are filled, or when neither player can move (either because he has no legal move or because his counters are gone). The winner is the person with the most pieces on the board.

Two examples will clarify the rules: In Figure 28, red can play only on cells 43, 44, 45 and 46. In each case he captures and reverses a single piece. In Figure 29, if red plays on cell 22 he is compelled to reverse six pieces: 21, 29, 36, 30, 38 and 46. As a result the board, which formerly was mostly black, suddenly becomes mostly red. Dramatic reversals of color are characteristic of this unusual game, and it is often difficult to say who has the better game until the last few plays are made. The player with the fewest pieces frequently has a strong positional advantage.

Some pointers for beginners: If possible, confine early play to the central sixteen squares, and try especially to occupy cells 19, 22, 43 and 46. The first player forced outside this area is usually placed at a disadvantage. Outside the central sixteen squares, the most valuable cells to occupy are the corners of the board. For this reason it is unwise to play on cells 10, 15, 50 or 55, because this gives your opponent a chance to take the corner cells. Next to the corners, the most desirable cells are those that are next but one to the corners (3, 6, 17, 24, 41, 48, 59 and 62). Avoid giving your opponent a chance to occupy these cells. Deeper rules of strategy will occur to any player who advances beyond the novice stage.

Little in the way of analysis has been published about reversi; it is hard to say who, if either player, has the advantage on even a board as small as four-by-four. Here is a problem some readers may enjoy trying to solve. Is it possible for a game to occur in which a player, before his tenth move, wins by removing *all* the enemy pieces from the board?

Two Englishmen, Lewis Waterman and John W. Mollett, both claimed to be the sole inventor of reversi. Each called the other a fraud. In the late 1880's, when the game was enormously popular in England, rival handbooks and rival firms for the manufacture of equipment were authorized by the two claimants. Regardless of who invented it, reversi is a game that combines complexity of structure with rules of delightful simplicity, and a game that does not deserve oblivion.

ADDENDUM

THE GAME of Maharajah (which I had found in R. C. Bell's *Board and Table Games*) can always be won by the player with conventional pieces if he plays circumspectly. Richard A. Blue, Dennis A. Keen, William Knight and Wallace Smith all sent strategies against which the maharajah could not save himself, but the most efficient line of play came from William E. Rudge, then a physics student at Yale University. If Rudge's strategy is bug-free, as it seems to be, the maharajah can always be captured in 25 moves or less.

The strategy is independent of the moves made by M (the maharajah) except for three possible moves. Only the moves of the offense are listed:

1. P — QR4
2. P — QR5
3. P — QR6
4. P — QR7
5. P — K3
6. N — KR3
7. N — KB4
8. B — Q3
9. Castles
10. Q — KR5
11. N — QB3
12. QN — Q5
13. R — QR6
14. P — QN4

M is now forced to move to his first or second row.

15. P — KR3

This move is made only if M is on his KN2. The move forces M to leave the corner-to-corner diagonal, permitting the following move.

16. B — QN2
17. R — QR1
18. R — K6
19. KR — QR6
20. R — K7

M is forced to retreat to his first row.

21. KR — K6
22. B — KN7

This move need be made only if M is on his KB1 or KN1.

23. P — QB3

This move is made only if M is on his KN1.

24. Q — K8

The maharajah can now be captured on the next move.

Moves 1 through 4 may be interchanged with moves 5 through 9, provided the sequence in each group is maintained. This interchange may be necessary if M blocks a pawn. Moves 15 and 22 are stalling moves, required only when M is on the squares indicated. Move 23 is required only if M must be forced over to the queen's side of the board.

Not much is known about the early history of reversi. It seems to have first appeared in London in 1870 as "The Game of Annexation," played on a cross-shaped board. A second version, using the standard eight-by-eight checkerboard, was called "Annex, a Game of Reverses." By 1888 the name had become reversi, and the game was something of a fad in England. Articles about it ran in a London newspaper called *The Queen* in the spring of 1888. Later, an elaboration called "Royal Reversi," using cubes with differently colored sides, was manufactured by the London firm of Jacques & Son. (For a description of Royal Reversi and a picture of the board, see *The Book of Table Games,* by "Professor Hoffman" [Angelo Lewis], pages 621–623.)

Reversi, and games derived from it, have been sold in more recent years, in the United States, under a variety of names. In 1938 Milton Bradley introduced Chameleon, a variant of Royal Reversi. Tryne Products brought out reversi, about 1960, as a game called "Las Vegas Backfire." Exit, a game that appeared in England in 1965, is reversi played on a board with circular cells. A fixed cover for each cell can be turned to make the cell red, blue or white (neutral), thus eliminating the need for pieces.

ANSWERS

CAN A REVERSI PLAYER, in less than ten moves, win a game by eliminating every enemy piece? The answer is yes. In my *Scientific American* column I gave what I believed then to be the shortest possible reversi game (corresponding to the "fool's mate" of chess), the first player winning on his eighth move. (I had found the game in an old reversi handbook.) But two readers discovered shorter games.

D. H. Peregrine, of Jesus College, Oxford, sent the following six-mover:

First Player	*Second Player*
28	29
36	37
38	45
54	35
34	27
20	

And Jon Petersen, Menlo Park, California, sent this slightly different six-move win:

First Player	*Second Player*
36	28
37	29
21	30
39	44
35	45
53	

CHAPTER SEVEN

□

Packing Spheres

SPHERES OF identical size can be piled and packed together in many different ways, some of which have fascinating recreational features. These features can be understood without models, but if the reader can obtain a supply of 30 or more spheres, he will find them an excellent aid to understanding. Table-tennis balls are perhaps the best for this purpose. They can be coated with rubber cement, allowed to dry, then stuck together to make rigid models.

First let us make a brief two-dimensional foray. If we arrange spheres in square formation [*see Fig. 30, right*], the number of balls involved will of course be a square number. If we form a triangle [*see Fig. 30, left*], the number of balls is a triangular number. These are the simplest examples of what the ancients

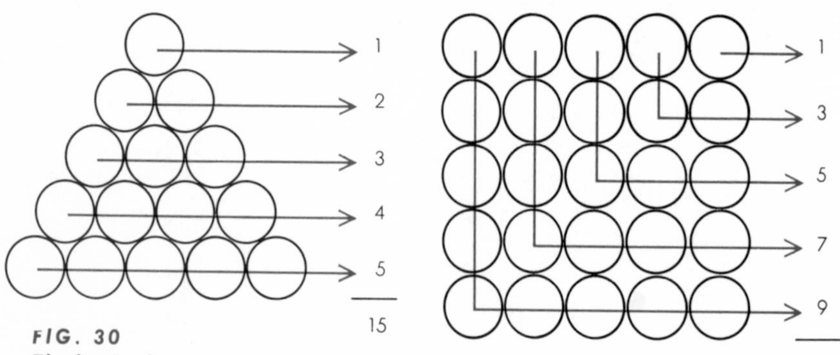

FIG. 30
The basis of triangular numbers (left) and of square numbers (right).

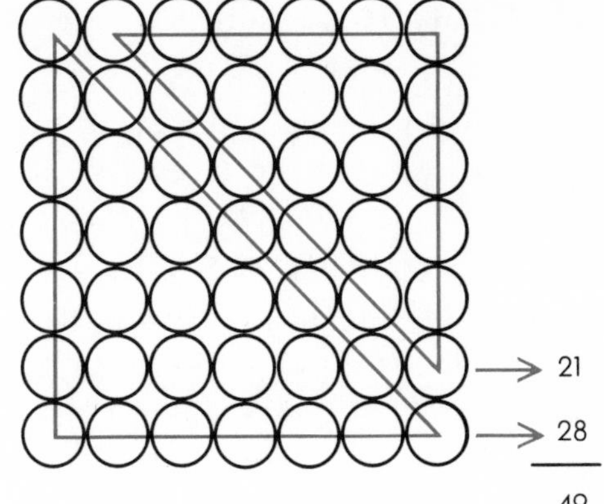

FIG. 31
Square and triangular numbers are related.

called "figurate numbers." They were intensively studied by early mathematicians (a famous treatise on them was written by Blaise Pascal), and although little attention is paid them today, they still provide intuitive insights into many aspects of elementary number theory.

For example, it takes only a glance at Figure 30, left, to see that the sum of any number of consecutive positive integers, beginning with 1, is a triangular number. A glance at Figure 30, right, shows that square numbers are formed by the addition of consecutive *odd* integers, beginning with 1. Figure 31 makes immediately evident an interesting theorem known to the ancient Pythagoreans: Every square number is the sum of two consecutive triangular numbers. The algebraic proof is simple. A triangular number with n units to a side is the sum of $1 + 2 + 3 + \ldots n$, and can be expressed by the formula $\frac{1}{2}n(n + 1)$. The preceding triangular number has the formula $\frac{1}{2}n(n - 1)$. If we add the two formulas and simplify, the result is n^2. Are there numbers that are simultaneously square and triangular? Yes, there are infinitely many of them. The smallest (not counting 1, which belongs to any figurate series) is 36; then the series continues: 1225, 41616, 1413721, 48024900 . . . It is not so easy to devise a formula for the nth term of this series.

Three-dimensional analogies of the plane-figurate numbers are obtained by piling spheres in pyramids. Three-sided pyramids, the base and sides of which are equilateral triangles, are models of what are called the tetrahedral numbers. They form the series

1, 4, 10, 20, 35, 56, 84 . . . and can be represented by the formula $\frac{1}{6}n(n + 1)(n + 2)$, where n is the number of balls along an edge. Four-sided pyramids, with square bases and equilateral triangles for sides (*i.e.*, half of a regular octahedron), represent the (square) pyramidal numbers 1, 5, 14, 30, 55, 91, 140 . . . They have the formula $\frac{1}{6}n(n + 1)(2n + 1)$. Just as a square can be divided by a straight line into two consecutive triangles, so can a square pyramid be divided by a plane into two consecutive tetrahedral pyramids. (If you build a model of a pyramidal number, the bottom layer has to be kept from rolling apart. This can be done by placing rulers or other strips of wood along the sides.)

Many old puzzles exploit the properties of these two types of pyramidal number. For example, in making a courthouse monument out of cannon balls, what is the smallest number of balls that can first be arranged on the ground as a square, then piled in a square pyramid? The surprising thing about the answer (4,900) is that it is the *only* answer. (The proof of this is difficult, and was not achieved until 1918.) Another example: A grocer is displaying oranges in two tetrahedral pyramids. By putting together the oranges in both pyramids he is able to make one large tetrahedral pyramid. What is the smallest number of oranges he can have? If the two small pyramids are the same size, the unique answer is 20. If they are different sizes, what is the answer?

Imagine now that we have a very large box, say a crate for a piano, which we wish to fill with as many golf balls as we can. What packing procedure should we use? First we form a layer packed as shown by the unshaded circles with light gray circumferences in Figure 32. The second layer is formed by placing balls in alternate hollows as indicated by the shaded circles with black rims. In making the third layer we have a choice of two different procedures:

1. We place each ball on a hollow A that is directly above a ball in the first layer. If we continue in this way, placing the balls of each layer directly over those in the next layer but one, we produce a structure called hexagonal close-packing.

2. We place each ball in a hollow B, directly above a hollow in the first layer. If we follow this procedure for each layer (each ball will be directly above a ball in the third layer beneath it), the result is known as cubic close-packing. Both the square and

the tetrahedral pyramids have a packing structure of this type, though on a square pyramid the layers run parallel to the sides rather than to the base.

In forming the layers of a close-packing we can switch back and forth whenever we please from hexagonal to cubic packing to produce various hybrid forms of close-packing. In all these forms — cubic, hexagonal and hybrid — each ball touches twelve other balls that surround it, and the density of the packing (the ratio of the volume of the spheres to the total space) is $\pi/\sqrt{18} = .74048+$, or almost 75 per cent.

Is this the largest density obtainable? No denser packing is known, but in an article published in 1958 (on the relation of close-packing to froth) H. S. M. Coxeter, of the University of Toronto, made the startling suggestion that perhaps the densest packing has not yet been found. It is true that no more than twelve balls can be placed so that all of them touch a central sphere, but a thirteenth ball can *almost* be added. The large leeway here in the spacing of the twelve balls, in contrast to the complete absence of leeway in the close-packing of circles on a plane, suggests that there might be some form of irregular packing that would be denser than .74. No one has yet proved that no denser packing is possible, or even that twelve point-contacts for each sphere are necessary for densest packing. As a result of

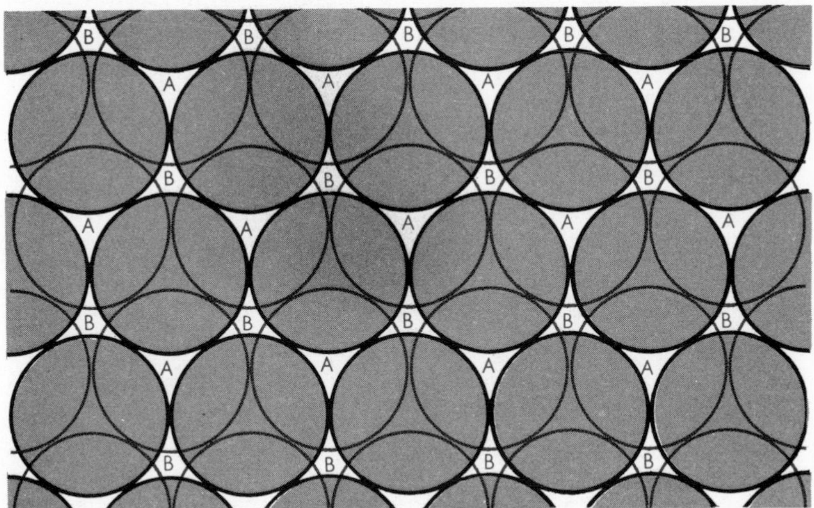

FIG. 32
In hexagonal close-packing, balls go in hollows labeled A; in cubic, in hollows labeled B.

Coxeter's conjecture, George D. Scott, of the University of Toronto, recently made some experiments in random packing by pouring large numbers of steel balls into spherical flasks, then weighing them to obtain the density. He found that stable random-packings had a density that varied from about .59 to .63. So if there is a packing denser than .74, it will have to be carefully constructed on a pattern that no one has yet thought of.

Assuming that close-packing is the closest packing, readers may like to test their packing prowess on this exceedingly tricky little problem. The interior of a rectangular box is ten inches on each side and five inches deep. What is the largest number of steel spheres one inch in diameter that can be packed in this space?

If close-packed circles on a plane expand uniformly until they fill the interstices between them, the result is the familiar hexagonal tiling of bathroom floors. (This explains why the pattern is so common in nature: the honeycomb of bees, a froth of bubbles between two flat surfaces almost in contact, pigments in the retina, the surface of certain diatoms and so on.) What happens when closely packed spheres expand uniformly in a closed vessel, or are subjected to uniform pressure from without? Each sphere becomes a polyhedron, its faces corresponding to planes that were tangent to its points of contact with other spheres. Cubic close-packing transforms each sphere into a rhombic dodecahedron [*see Fig. 33, top*], the twelve sides of which are congruent rhombi. Hexagonal close-packing turns each ball into a trapezo-rhombic dodecahedron [*see Fig. 33, bottom*], six faces of which are rhombic and six trapezoidal. If this figure is sliced in half along the gray plane and one half is rotated 60 degrees, it becomes a rhombic dodecahedron.

In 1727 the English physiologist Stephen Hales wrote in his book *Vegetable Staticks* that he had poured some fresh peas into a pot, compressed them and had obtained "pretty regular dodecahedrons." The experiment became known as the "peas of Buffon" (because the Comte de Buffon later wrote about a similar experiment), and most biologists accepted it without question until Edwin B. Matzke, a botanist at Columbia University, repeated the experiment. Because of the irregular sizes and shapes of peas,

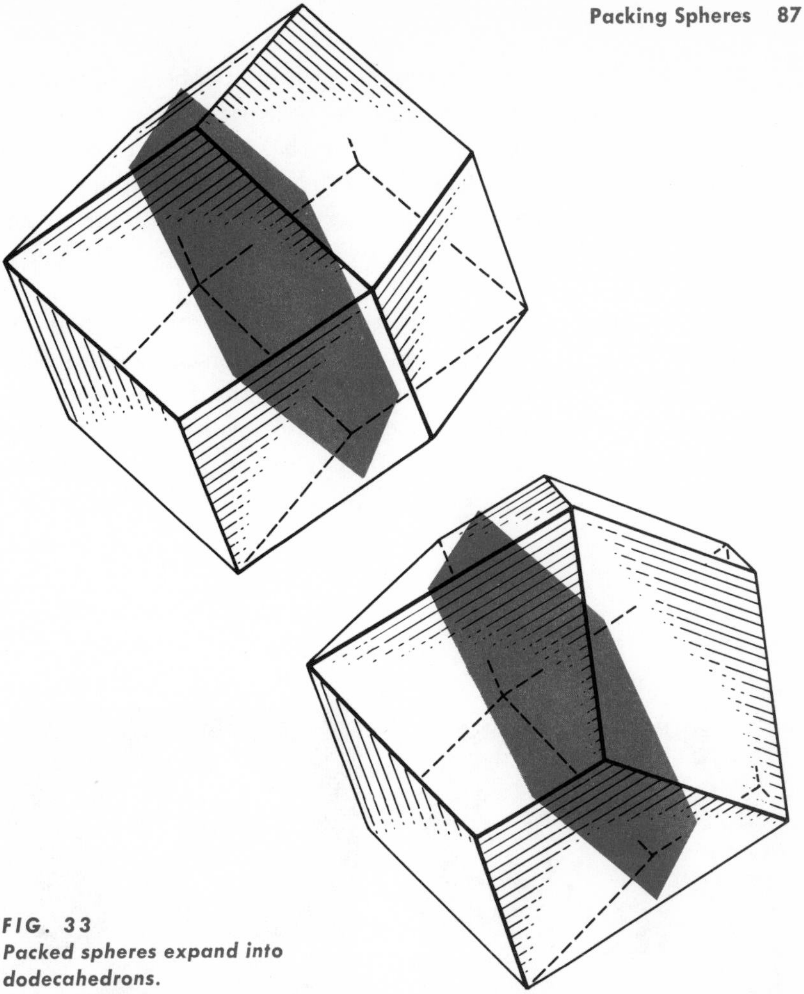

FIG. 33
Packed spheres expand into dodecahedrons.

their nonuniform consistency and the random packing that results when peas are poured into a container, the shapes of the peas after compression are too random to be identifiable. In experiments reported in 1939 Matzke compressed lead shot and found that if the spheres had been cubic close-packed, rhombic dodecahedrons were formed; but if they had been randomly packed, irregular fourteen-faced bodies predominated. These results have important bearing, Matzke has pointed out, on the study of such structures as foam, and living cells in undifferentiated tissues.

The problem of closest packing suggests the opposite question: What is the *loosest* packing; that is, what rigid structure will have the lowest possible density? For the structure to be rigid, each sphere must touch at least four others, and the contact points must not be all in one hemisphere or all on one equator of the sphere. In his *Geometry and the Imagination,* first published in Germany in 1932, David Hilbert describes what was then believed to be the loosest packing: a structure with a density of .123. In the following year, however, two Dutch mathematicians, Heinrich Heesch and Fritz Laves, published the details of a much looser packing with a density of only .0555 [*see Fig. 34*]. Whether there are still looser packings is another intriguing question that, like the question of the closest packing, remains undecided.

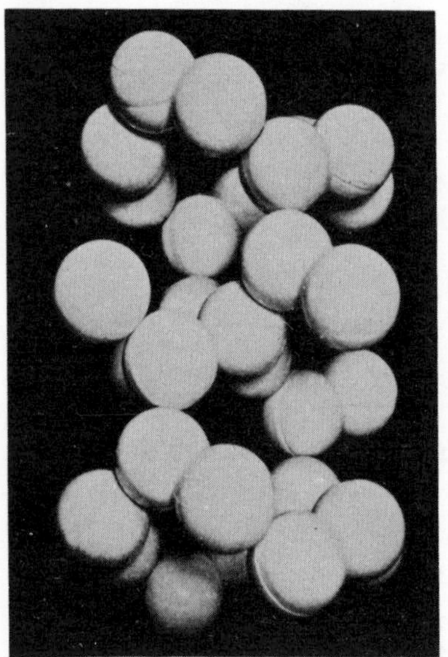

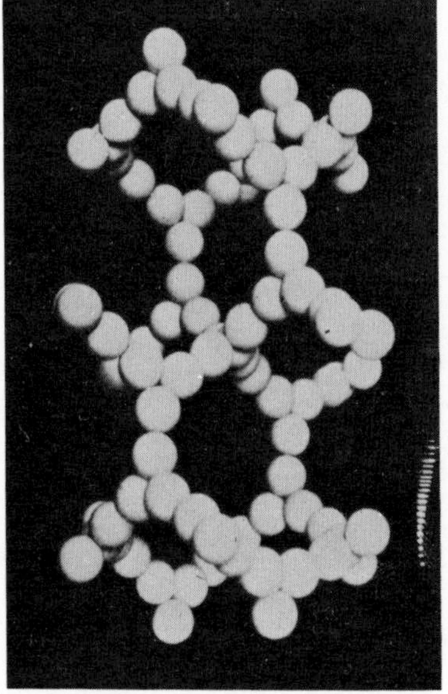

FIG. 34
The Heesch and Laves loose-packing. Large spheres are first packed as shown on left, then each sphere is replaced by three smaller spheres to obtain the packing shown on right. It has a density of .055+.

ADDENDUM

THE UNIQUE ANSWER of 4,900 for the number of balls that will form both a square and a square-based pyramid was proved by G. N. Watson in *Messenger of Mathematics,* new series, Vol. 48, 1918, pages 1–22. This had been conjectured as early as 1875 by the French mathematician Edouard Lucas. Henry Ernest Dudeney makes the same guess in his answer to problem 138, *Amusements in Mathematics* (1917).

There is a large literature on numbers that are both triangular and square. Highlights are cited in an editorial note to problem E1473, *American Mathematical Monthly,* February 1962, page 169, and the following formula for the nth square triangular number is given:

$$\frac{(17 + 12\sqrt{2})^n + (17 - 12\sqrt{2})^n - 2}{32}$$

The question of the densest possible regular packing of spheres has been solved for all spaces up through eight dimensions. (See *Proceedings of Symposia in Pure Mathematics,* Vol. 7, American Mathematical Society, 1963, pages 53–71.) In 3-space, the question is answered by the regular close-packings described earlier, which have a density of .74+. But, as Constance Reid notes in her *Introduction to Higher Mathematics* (1959), when 9-space is considered, the problem takes one of those sudden, mysterious turns that so often occur in the geometries of higher Euclidean spaces. So far as I know, no one yet knows how to regularly close-pack hyperspheres in 9-space.

Nine-space is also the turning point for the related problem of how many congruent spheres can be made to touch another sphere of the same size. It was not until 1953 that K. Schütte and B. L. van der Waerden (in *Das Problem der dreizehn Kugeln, Math. Ann.,* Vol. 125, 1953, pages 325–334) first proved that the answer in 3-space is 12. (For a later proof, see "The Problem of the 13 Spheres" by John Leech, in *Mathematical Gazette,* Vol. 40, No. 331, February 1956, pages 22–23.) The corresponding problem on the plane has the obvious answer of 6 (no more than six pennies can touch another penny), and if we think of a straight line as a degenerate "sphere," the answer for 1-space is

2. In four-dimensions it has been proved that 24 hyperspheres can touch a 25th sphere, and for spaces of 5, 6, 7 and 8 dimensions, the maximum number of hyperspheres is known to be 40, 72, 126 and 240 respectively. But in 9-space, the problem remains unsolved.

ANSWERS

THE SMALLEST NUMBER of oranges that will form two tetrahedral pyramids of different sizes, and also one larger tetrahedral pyramid, is 680. This is a tetrahedral number that can be split into two smaller tetrahedral numbers: 120 and 560. The edges of the three pyramids are 8, 14 and 15.

A box ten inches square and five inches deep can be close-packed with one-inch-diameter steel balls in a surprising variety of ways, each accommodating a different number of balls. The maximum number, 594, is obtained as follows: Turn the box on its side and form the first layer by making a row of five, then a row of four, then of five, and so on. It is possible to make eleven rows (six rows of five each, five rows of four each), accommodating 50 balls and leaving a space of more than .3 inch to spare. The second layer also will take eleven rows, alternating four and five balls to a row, but this time the layer begins and ends with four-ball rows, so that the number of balls in the layer is only 49. (The last row of four balls will project .28+ inch beyond the edge of the first layer, but because this is less than .3 inch, there is space for it.) Twelve layers (with a total height of 9.98+ inches) can be placed in the box, alternating layers of 50 balls with layers of 49, to make a grand total of 594 balls.

CHAPTER EIGHT

□

The Transcendental Number Pi

Pi's face was masked, and it was understood that none could behold it and live. But piercing eyes looked out from the mask, inexorable, cold, and enigmatic.

— Bertrand Russell,
"The Mathematician's Nightmare,"
in *Nightmares of Eminent Persons*

THE RATIO of a circle's circumference to its diameter, symbolized by the Greek letter pi, pops up in all sorts of places that have nothing to do with circles. The English mathematician Augustus de Morgan once wrote of pi as "this mysterious 3.14159 . . . which comes in at every door and window, and down every chimney." To give one example, if two numbers are picked at random from the set of positive integers, what is the probability that they will have no common divisor? The surprising answer is six divided by the square of pi. It is pi's connection with the circle, however, that has made it the most familiar member of the infinite class of transcendental numbers.

What is a transcendental number? It is described as an irrational number that is not the root of an algebraic equation that has rational coefficients. The square root of two is irrational, but it is an "algebraic irrational" because it is a root of the equation $x^2 = 2$. Pi cannot be expressed as the root of such an equation, but only as the limit of some type of infinite process. The decimal form of pi, like that of all irrational numbers, is endless and nonrepeating.

No fraction, with integers above and below the line, can exactly equal pi, but there are many simple fractions that come amazingly close. The most remarkable was recorded in the fifth century A.D. by Tsu Ch'ung-Chih, a famous Chinese astronomer, and was not discovered in the Occident until 1,000 years later. We can obtain this fraction by a kind of numerological hocus-pocus. Write the first three odd integers in pairs: 1, 1, 3, 3, 5, 5; then put the last three above the first three to make the fraction 355/113. It is hard to believe, but this gives pi to an accuracy of six decimal places. There are also roots that come close to pi. The square root of 10 (3.162 . . .) was widely used for pi in ancient times, but the cube root of 31 (3.1413 . . .) is much closer. (More numerology: 31 comprises the first two digits of pi.) A cube with a volume of 31 cubic inches would have an edge that differed from pi by less than a thousandth of an inch. And the sum of the square root of 2 and the square root of 3 is 3.146+, also not a bad approximation.

Early attempts to find an exact value for pi were closely linked with attempts to solve the classic problem of squaring the circle. Is it possible to construct a square, using only a compass and a straightedge, that is exactly equal in area to the area of a given circle? If pi could be expressed as a rational fraction or as the root of a first- or second-degree equation, then it would be possible, with compass and straightedge, to construct a straight line exactly equal to the circumference of a circle. The squaring of the circle would quickly follow. We have only to construct a rectangle with one side equal to the circle's radius and the other equal to half the circumference. This rectangle has an area equal to that of the circle, and there are simple procedures for converting the rectangle to a square of the same area. Conversely, if the circle could be squared, a means would exist for constructing a line segment exactly equal to pi. However, there are ironclad proofs that pi is transcendental and that no straight line of transcendental length can be constructed with compass and straightedge.

There are hundreds of approximate constructions of pi, of which one of the most accurate is based on the Chinese astronomer's fraction mentioned earlier. In a quadrant of unit radius draw the lines shown in Figure 35 so that bc is 7/8 of the radius, dg is 1/2, de is parallel to ac, and df is parallel to be. The distance fg is easily shown to be 16/113 or .1415929+. Since 355/113 is

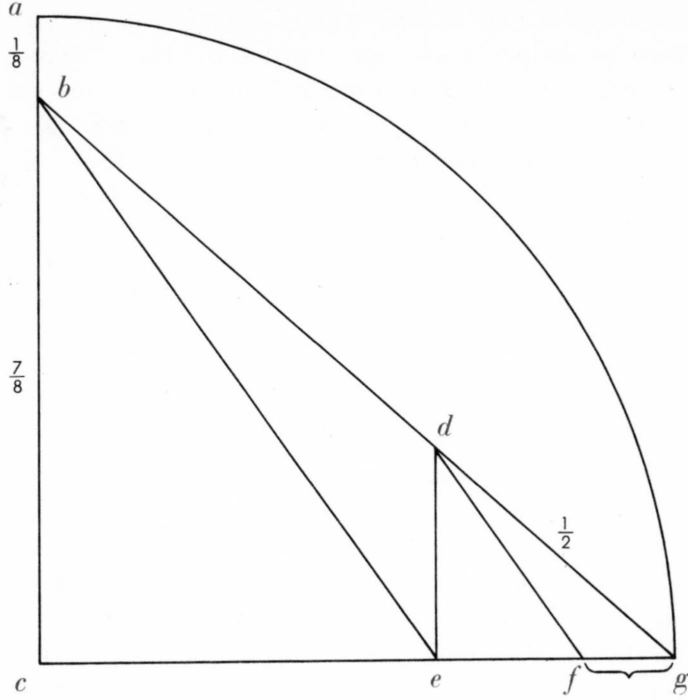

FIG. 35
How to construct a straight line with a length that differs from pi by less than .0000003.

3 + 16/113, we draw a line that is three times the radius, extend it by the distance *fg*, and we have a line differing from pi by less than a millionth of a unit.

Circle squarers who thought they had discovered an exact value for pi are legion, but none has excelled the English philosopher Thomas Hobbes in combining height of intellect with depth of ignorance. Educated Englishmen were not taught mathematics in Hobbes's day, and it was not until he was 40 that he looked into Euclid. When he read a statement of the Pythagorean theorem, he first exclaimed: "By God, this is impossible!" Then he threaded his way backward through the proof until he became convinced. For the rest of his long life Hobbes pursued geometry with all the ardor of a man in love. "Geometry hath in it something like wine," he later wrote, and it is said that he was accustomed, when better surfaces were wanting, to drawing geometrical figures on his thighs and bedsheets.

Had Hobbes been content to remain an amateur mathematician, his later years would have been more tranquil, but his monstrous egotism led him to think himself capable of great mathematical discoveries. In 1655, at the age of 67, he published in Latin a book titled *De corpore* (*Concerning Body*) that included an ingenious method of squaring the circle. The method was an excellent approximation, but Hobbes believed that it was exact. John Wallis, a distinguished English mathematician and cryptographer, exposed Hobbes's errors in a pamphlet, and thus began one of the longest, funniest and most profitless verbal duels ever to engage two brilliant minds. It lasted almost a quarter of a century, each man writing with skillful sarcasm and barbed invective. Wallis kept it up partly for his own amusement, but mainly because it was a way of making Hobbes appear ridiculous and thus casting doubt on his religious and political opinions, which Wallis detested.

Hobbes responded to Wallis' first attack by reprinting his book in English with an addition called *Six Lessons to the Professors of Mathematics....* (I trust the reader will forgive me if I shorten the endless 17th-century titles.) Wallis replied with *Due Correction for Mr. Hobbes in School Discipline for not saying his Lessons right.* Hobbes countered with *Marks of the Absurd Geometry, Rural Language, Scottish Church Politics, and Barbarisms of John Wallis;* Wallis fired back with *Hobbiani Puncti Dispunctio! or the Undoing of Mr. Hobbes's Points.* Several pamphlets later (meanwhile Hobbes had anonymously published in Paris an absurd method of duplicating the cube) Hobbes wrote: "I alone am mad, or they [the professors of mathematics] are all out of their senses: so that no third opinion can be taken, unless any will say that we are all mad."

"It needs no refutation," was Wallis' answer. "For if he be mad, he is not likely to be convinced by reason; on the other hand, if we be mad, we are in no position to attempt it."

The battle continued, with momentary periods of cease-fire, until Hobbes's death at the age of 91. "Mr. Hobbes has been always far from provoking any man," Hobbes wrote in one of his later attacks on Wallis [as a matter of fact, in social relations Hobbes was excessively timid], "though, when he is provoked, you find his pen as sharp as yours. All you have said is error and railing; that is, stinking wind, such as a jade lets fly when he is too hard girt upon a full belly. I have done. I have considered you now, but will not again. . . ."

This is not the place to go into details about Hobbes's curious "incapacity," as Wallis phrased it, "to be taught what he doth not know." Altogether, Hobbes published about a dozen different methods of squaring the circle. His first, and one of his best, is shown in Figure 36. Inside a unit square, draw arcs AC and BD. These are quarter arcs of circles with unit radii. Bisect arc BF at Q. Draw line RQ parallel with the side of the square and extend it so QS equals RQ. Draw line FS, extending it until it meets the side of square at T. BT, Hobbes asserted, is exactly equal to arc BF. Since arc BF is 1/12 the circumference of a circle with unit radius, pi will be six times the length of BT. This gives pi a value of 3.1419+.

One of the philosopher's major difficulties was his inability to believe that points, lines and surfaces could be regarded in the abstract as having less than three dimensions. "He seems to have gone down to the grave," writes Isaac Disraeli in his *Quarrels of*

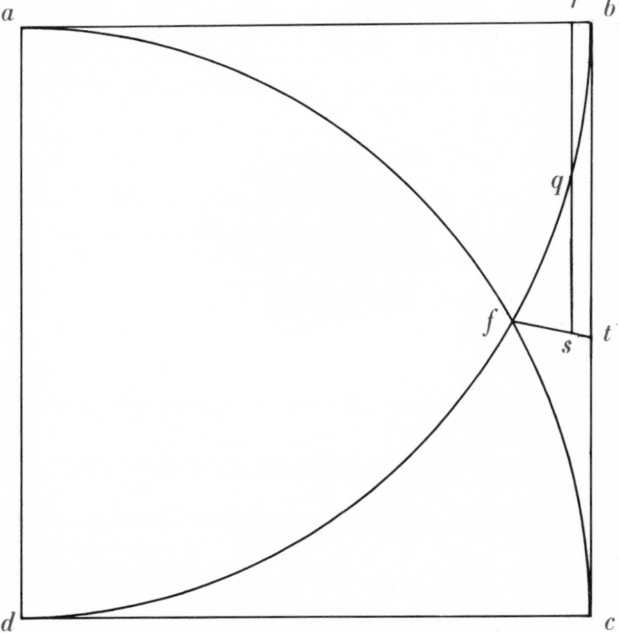

FIG. 36
Hobbes's First Method of squaring the circle.

Quadratura Circuli,
Cubatio Sphæræ,
Duplicatio Cubi,

Breviter demonſtrata.

Auct. Tho. Hobbes.

LONDINI:

Excudebat *J. C.* Sumptibus *Andrææ Crooke.* 1669.

No. 67

Title page of one of Hobbes's books that contains a method of circle squaring.

Authors, "in spite of all the reasonings of the geometricians on this side of it, with a firm conviction that its superficies had both depth and thickness." Hobbes presents a classic case of a man of genius who ventures into a branch of science for which he is ill prepared and dissipates his great energies on pseudo-scientific nonsense.

Although the circle cannot be squared, figures bounded by circular arcs often can be; this fact still arouses false hopes in many a circle squarer. An interesting example is shown in Figure 37.

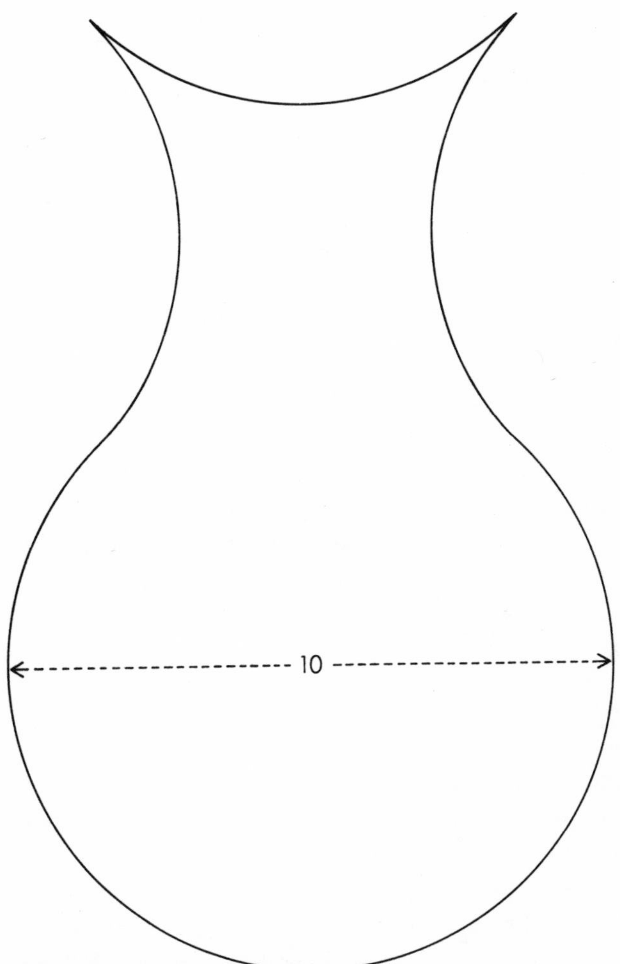

FIG. 37
How many square units does this figure contain?

The lower part of this vase is three-quarters of the circumference of a circle with a diameter of, say, ten inches. The upper half is bounded by three quarter-arcs of a circle the same size. How quickly can the reader give, down to the last decimal, the exact length of the side of a square that has the same area as this figure?

Close cousins to the circle squarers have been the pi computers; men who devoted years to computing by hand the decimals of pi beyond all previous computations. This can be done, of course, by using any infinite expression that converges on pi. Wallis himself discovered one of the simplest:

$$\pi = 2\left(\frac{2}{1} \times \frac{2}{3} \times \frac{4}{3} \times \frac{4}{5} \times \frac{6}{5} \times \frac{6}{7} \times \frac{8}{7} \times \frac{8}{9} \cdots\right)$$

The upper terms of these fractions are even numbers in sequence, taken in pairs. (Note the fortuitous resemblance of the first five lower terms to the digits in the Chinese astronomer's fraction!) A few decades later the German philosopher Gottfried Wilhelm von Leibniz found another beautiful formula:

$$\pi = 4\left(\frac{1}{1} - \frac{1}{3} + \frac{1}{5} - \frac{1}{7} + \frac{1}{9} \cdots\right)$$

The most indefatigable of pi computers was the English mathematician William Shanks. Over a 20-year period he managed to calculate pi to 707 decimals. Alas, poor Shanks made an error on his 528th decimal, and all the rest are wrong. (This was not discovered until 1945, so Shanks's 707 decimals are still found in many current books.) In 1949 the electronic computer ENIAC was used for 70 machine hours to calculate pi to more than 2,000 decimals; later another computer carried it to more than 3,000 decimals in 13 minutes. By 1959, a computer in England and another in France had computed pi to 10,000 decimal places.

One of the strangest aspects of Shanks's 707 decimals was the fact that they seemed to snub the number 7. Each digit appeared about 70 times in the first 700 decimals, just as it should, except 7, which appeared a mere 51 times. "If the cyclometers and the apocalyptics would lay their heads together," wrote De Morgan, "until they came to a unanimous verdict on this phenomenon, and would publish nothing until they are of one mind, they would earn the gratitude of their race." I hasten to add that the corrected value of pi to 700 places restored the missing 7's. The intuitionist school of mathematics, which maintains that you cannot say of a statement that it is "either true or false" unless there is a known way by which it can be both verified and refuted, has always used as its stock example: "There are three consecutive 7's in pi." This must now be changed to five 7's. The new figures for pi show not only the expected number of triplets for each digit, but also several runs of 7777 (and one unexpected 999999).

So far pi has passed all statistical tests for randomness. This is disconcerting to those who feel that a curve so simple and beautiful as the circle should have a less-disheveled ratio between the way around and the way across, but most mathematicians believe that no pattern or order of any sort will ever be found in pi's decimal expansion. Of course the digits are not random in the sense that they represent pi, but then in this sense neither are the million random digits that have been published by the Rand Corporation of California. They too represent a single number, and an integer at that.

If it is true that the digits in pi are random, perhaps we are justified in stating a paradox somewhat similar to the assertion that if a group of monkeys pound long enough on typewriters, they will eventually type all the plays of Shakespeare. Stephen Barr has pointed out that if you set no limit to the accuracy with which two bars can be constructed and measured, then those two bars, without any markings on them, can communicate the entire *Encyclopaedia Britannica*. One bar is taken as unity. The other differs from unity by a fraction that is expressed as a very long decimal. This decimal codes the *Britannica* by the simple process of assigning a different number (excluding zero as a digit in the number) to every word and mark of punctuation in the language. Zero is used to separate the code numbers. Obviously the entire *Britannica* can now be coded as a single, but almost inconceivably long, number. Put a decimal point in front of this number, add 1, and you have the length of the second of Barr's bars.

Where does pi come in? Well, if the digits in pi are really random, then somewhere in this infinite pie there should be a slice that contains the *Britannica;* or, for that matter, any book that has been written, will be written, or could be written.

ADDENDUM

ON JULY 29, 1961, a year after the preceding chapter appeared in *Scientific American,* pi was carried to 100,265 decimal places by an IBM 7090 system at the IBM Data Center in New York. The work was done by Daniel Shanks (no relation to William Shanks; just another of those strange numerological coincidences that dog the history of pi) and John W. Wrench, Jr. The running time was one minute more than eight hours, then an additional 42 minutes were required to get the binary results into decimal form. Computing pi to a few thousand decimals is now a popular device for testing a new computer or training new programmers. "The mysterious and wonderful pi," writes Philip J. Davis (in his book *The Lore of Large Numbers*), "is reduced to a gargle that helps computing machines clear their throats."

It will probably not be long until pi is known to a million decimals. In anticipation of this, Dr. Matrix, the famous numerologist, has sent me a letter asking that I put on record his prediction that the millionth digit of pi will be found to be 5. His calculation is based on the third book of the King James Bible, chapter 14, verse 16 (it mentions the number 7, and the seventh word has five letters), combined with some obscure calculations involving Euler's constant and the transcendental number *e*.

Norman Gridgeman, of Ottawa, wrote to point out that Barr's bars can be reduced to a single bar with a scratch on it. The scratch divides the bar into two lengths, the ratio of which codes the *Britannica* in the manner previously described.

ANSWERS

IT WAS suggested that the reader give the side of a square equal in area to the vase-shaped figure in Figure 38, bounded by arcs of a circle with a diameter of ten inches. The answer is also ten

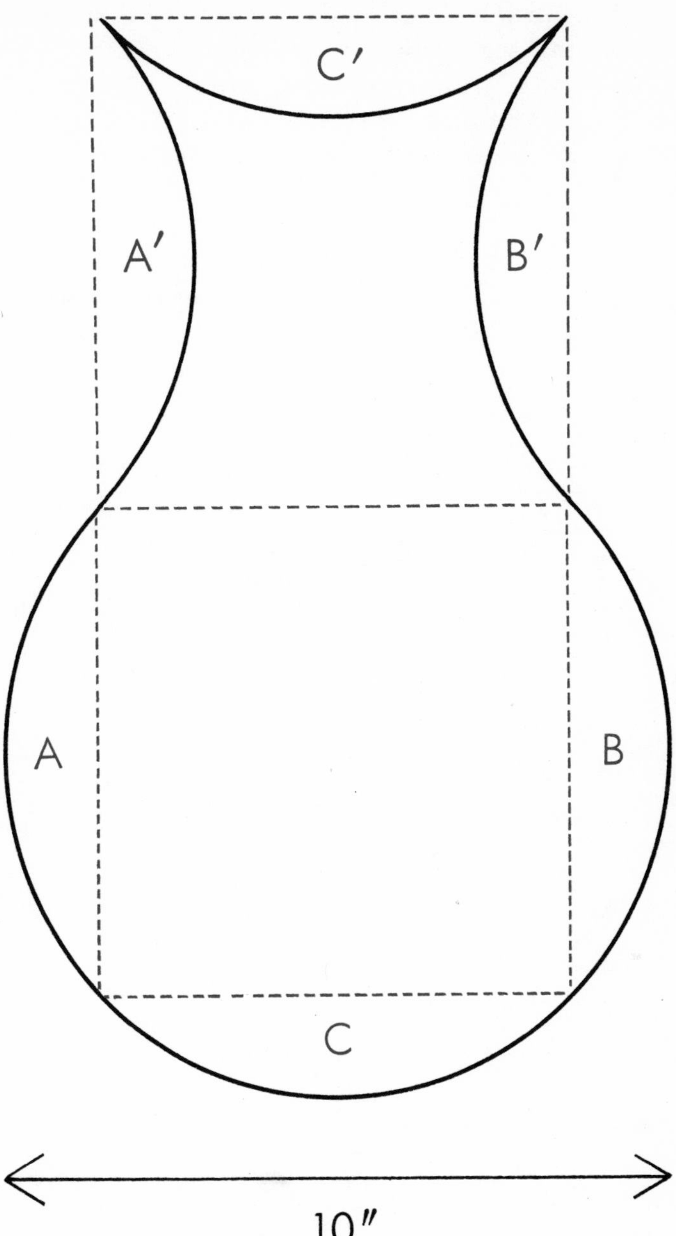

FIG. 38
How to square the vase.

inches. If we draw the broken squares shown in the illustration, it is obvious that segments A, B, C will fit into spaces A′, B′, C′ to form two squares with a combined area of 100 square inches. Figure 39 shows how the vase can be "squared" by cutting it into as few as three parts that will form a ten-inch square.

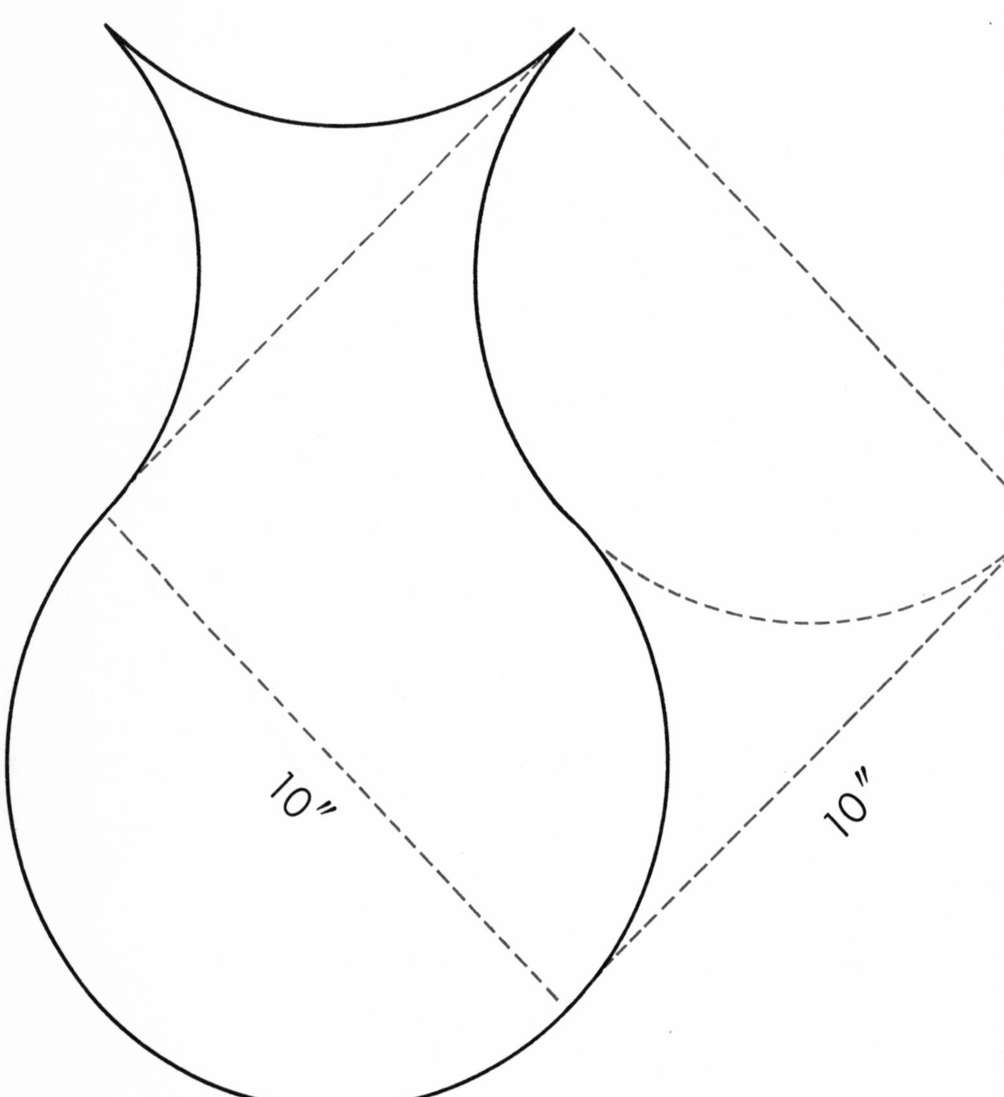

FIG. 39
Three-piece vase-to-square.

CHAPTER NINE

□

Victor Eigen: Mathemagician

Luzhin had no difficulty in learning several card tricks. . . . He found a mysterious pleasure, a vague promise of still unfathomed delights, in the crafty and accurate way a trick would come out. . . .

— Vladimir Nabokov, *The Defense*

AN INCREASING number of mathematically inclined amateur conjurers have lately been turning their attention toward "mathemagic": tricks that rely heavily on mathematical principles. Professional magicians shy away from such tricks because they are too cerebral and boring for most audiences, but as parlor stunts presented more in the spirit of puzzles than of feats of magic, they can be interesting and entertaining. My friend Victor Eigen, an electronics engineer and past president of the Brotherhood of American Wand Wielders, manages to keep posted on the latest developments in this curious field, and it was in the hope of finding some off-beat material for this department that I paid him a visit.

The front door was opened by Victor — a plump, gray-haired man in his mid-fifties with humorous creases around his eyes. "Do you mind sitting in the kitchen?" he asked as he led me toward the back of his apartment. "My wife's absorbed in a television program and I think we'd best not disturb her until it's over. How do you want your bourbon?"

We sat on opposite sides of the kitchen table and clinked glasses. "To mathemagic," I said. "What's new?"

Victor lost no time in taking a deck of cards from his shirt pocket. "The latest thing out in cards is the Gilbreath principle. It's a whimsical theorem discovered by Norman Gilbreath, a young California magician." As he talked, his short fingers skillfully arranged the deck so that red and black cards alternated throughout. "You know, I'm sure, that riffle shuffling is notoriously inefficient as a method of randomizing."

"No, I didn't realize that."

Victor's eyebrows went up. "Well, this ought to convince you. Please give the deck one thorough riffle shuffle."

I cut the deck into two parts and shuffled them together.

"Take a look at the faces," he said. "You'll see that the alternating color arrangement has been pretty well destroyed."

"Of course."

"Now give the deck a cut," he went on, "but cut between two cards of the same color. Square up the pack and hand it to me face down."

I did as he suggested. He held the deck under the table where it was out of sight for both of us. "I'm going to try to distinguish the colors by sense of touch," he said, "and bring out the cards in red-black pairs." Sure enough, the first pair he tossed on the table consisted of one red and one black card. The second pair likewise. He produced a dozen such pairs.

"But how . . . ?"

Victor interrupted with a laugh. He slapped the rest of the deck on the table and started taking cards from the top, two at a time, tossing them face up. Each pair contained a red and a black card. "Couldn't be simpler," he explained. "The shuffle and cut — remember, the cut must be between two cards of the same color — destroys the alternation of red and black all right, but it leaves the cards strongly ordered. Each pair still contains both colors."

"I can't believe it!"

"Well, think about it a bit and you'll see why it works, but it's not so easy to state a proof in a few words. By the way, my friend Edgar N. Gilbert, of Bell Telephone Laboratories, included an interesting puzzle along similar lines in a recent unpublished paper of his on card shuffling and information theory. Here, I've jotted it down for you."

He handed me a sheet on which was printed:

T L V E H E D I N S A G M E L R L I E N A T G O V R A R
G I A N E S T Y O F O F I F F O S H H R A V E M E V S O

"That's a garbled sentence," he said, "from a *Scientific American* article of five years ago. Gilbert wrote each letter on a card, then arranged the deck so it spelled the sentence from top down. He cut the cards into two piles, riffled them together, then copied down the new sequence of letters. It takes, he tells me, the average person about half an hour to unscramble them. The point is that one riffle shuffle is such a poor destroyer of information conveyed by the original sequence of cards, and the redundancy of various letter combinations in English is so high, that it's extremely unlikely — in fact, Gilbert computes the exact probability in his paper — that the message one finds is different from the correct one."

I rattled the ice cubes in my glass.

"Before we refill," Victor said, "let me show you an ingenious experiment in precognition. We'll need your glass and nine playing cards." He arranged nine cards, with values from one to nine, on the table in the form of the familiar three-by-three magic square [*see Fig. 40*]. The cards were all hearts, except for the five of spades in the center. He took an envelope from his pocket and placed it beside the square.

"I want you to put your glass on any one of the nine cards," he said, "but first let me explain that in this envelope is a file card on which I have jotted down some instructions. The instructions are based on my guesses as to the card you're going to choose, and how you are going to move the glass at random from card to card. If my guesses are correct, your glass will end on the card in the center." He tapped his finger on the five of spades. "Now put your glass on any card, including the center one if you wish."

I placed my glass on the two of hearts.

"Just as I expected," he chuckled. He took the file card from the envelope and held it so I could read the following instructions:

1. Take away the seven.
2. Move seven times and take away the eight.
3. Move four times, take away the two.
4. Move six times, take away the four.
5. Move five times, take the nine.
6. Move twice, take the three.
7. Move once, take the six.
8. Move seven times, take the ace.

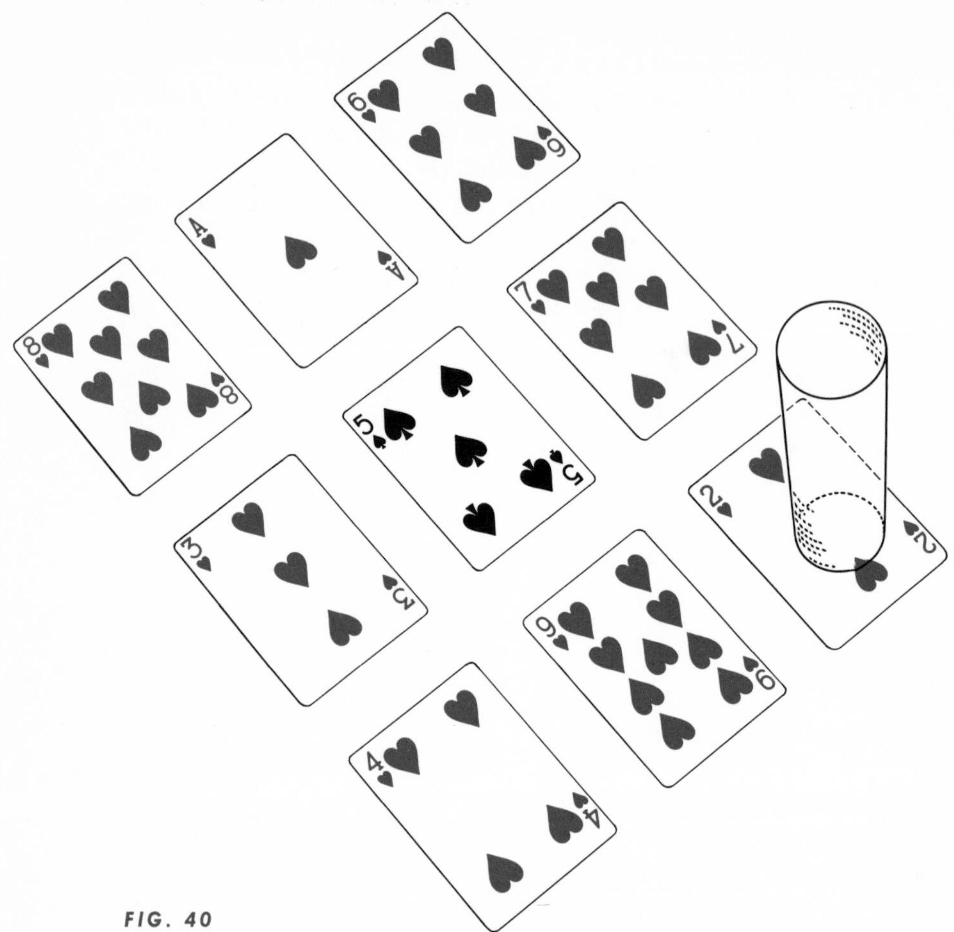

FIG. 40
Cards and a glass arranged for a demonstration of precognition.

A "move," he explained, consists of transferring the glass to an adjacent card above, below or on either side, but not diagonally. I followed the instructions carefully, making all moves as random as I could. To my vast surprise the glass never rested on a card that I was asked to remove, and after eight cards had been taken away, there was my glass, resting on the five of spades just as Victor had predicted!

"You've befuddled me completely," I admitted. "Suppose I had originally placed my glass on the seven of hearts, the first card removed?"

"I must confess," he said, "that a bit of nonmathematical chicanery is involved. The magic-square arrangement has nothing to do with the trick. Only the positions of the cards matter. Those in the odd positions — the four corners and the center — form one set; those in the even positions form a set of opposite parity. When I saw that you first placed your glass on a card in the odd set, I showed you the instructions you see here. If you had placed your glass on a card in the even set, I would have turned over the envelope before I took out the file card."

He flipped over the card. On its back was a second set of instructions. They read:

1. Take away the six.
2. Move four times and take away the two.
3. Move seven times, take away the ace.
4. Move three times, take away the four.
5. Move once, take the seven.
6. Move twice, take the nine.
7. Move five times, take the eight.
8. Move three times, take the three.

"You mean that these two sets of instructions — one to use if I start on an even-positioned card, and the other if I start on an odd — will always guide the glass to the center?"

Victor nodded. "Why don't you print both sides of the card in your department and let your readers figure out why the trick has to work?"

After refilling our glasses, Victor said: "Quite a number of ESP-type tricks exploit a parity principle. Here's one that seems to require clairvoyance." He handed me a blank sheet of paper and a pencil. "While my back is turned, I want you to draw a complicated closed curve that crosses itself at least a dozen times, but never more than once at any one point." He turned his chair so that he faced the wall while I drew the curve [*see Fig. 41*].

"Label each intersection with a different letter," he said over his shoulder.

I did as I was told.

"Now put your pencil on any spot along the curve and start tracing it. Each time you come to a crossing, call out the letter. Keep this up until you've traced the entire curve, but at some point along the way — it doesn't matter where — switch two

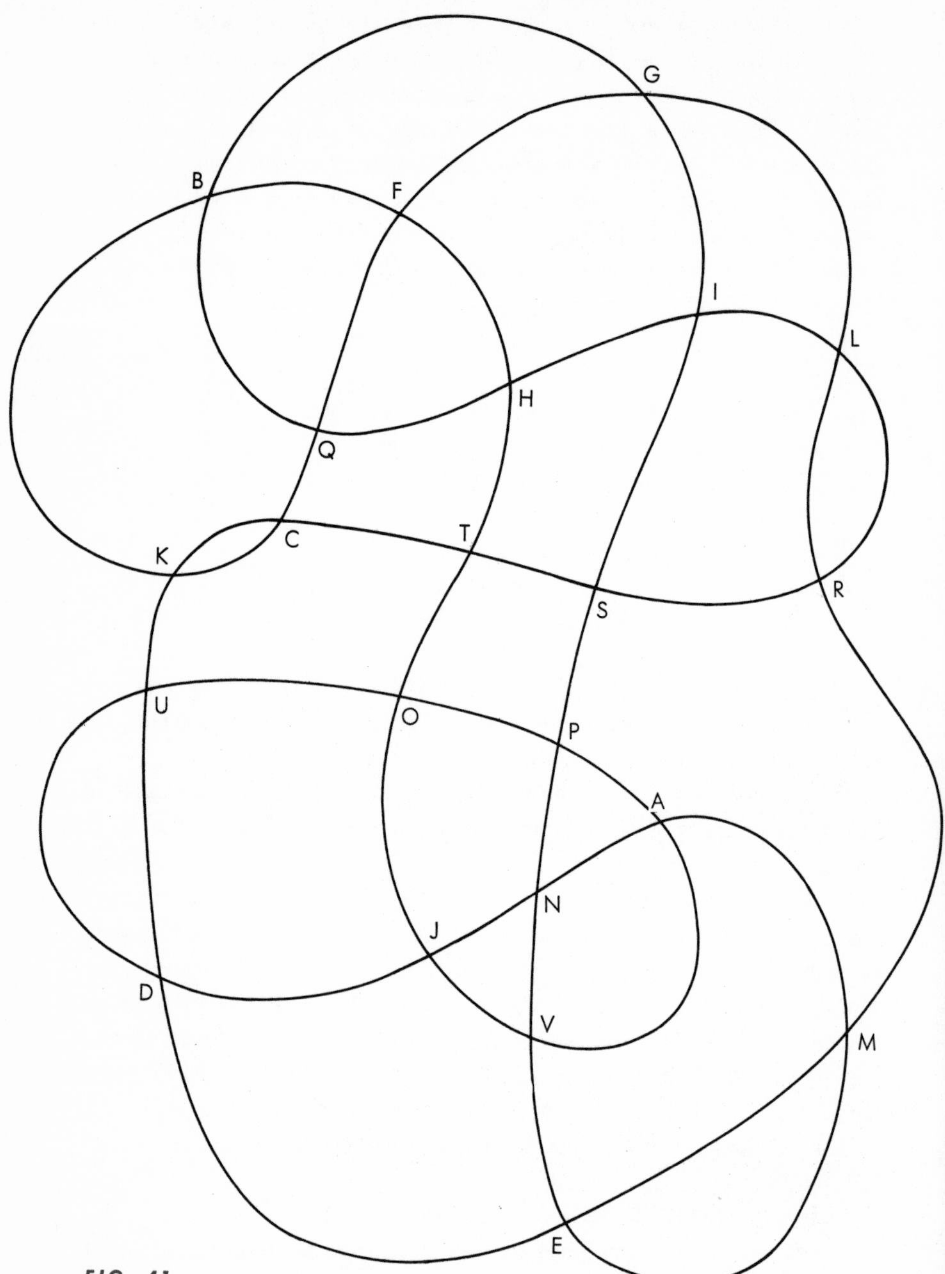

FIG. 41
Randomly drawn and labeled closed curve for an experiment in clairvoyance.

letters as you call them. The two letters must be adjacent along the path. Don't tell me when you switch them."

I started at point N, moved up to P and continued along the curve, calling out the letters as I came to them. I could see that Victor was jotting them down on a pad. When I approached B for the second time, I saw that the letter after it was F, so I called out F and then B. I made the switch without a break in the timing of my calls, so that Victor would have no clue as to which pair had been switched.

As soon as I finished he said: "You switched B and F."

"Amazing!" I said. "How did you know?"

Victor chuckled and turned back to face me. "The trick's based on a topological theorem that's important in knot theory," he said. "You'll find it neatly proved in Hans Rademacher and Otto Toeplitz's book *The Enjoyment of Mathematics*." He tossed over the pad on which he had jotted down the letters. They were printed alternately above and below a horizontal line like this:

N S G Q I R T K D M L F C F H O V P U J A E
___
P I B H L S C U E R G Q K B T J A O D N M V

"If no switch is made," he explained, "then every letter must appear once above and once below the line. All I have to do is look for a letter that appears twice above, and a letter that appears twice below. Those will be the two letters that are exchanged."

"Beautiful!" I said.

Victor opened a box of soda crackers, took out two and placed them on the table, one to his right and one to his left. On both crackers he drew an arrow pointing north [*see Fig. 42*]. He held the cracker on the left between his thumb and middle finger as shown, then with the tip of his right forefinger he pressed down on corner A to turn the cracker over. It rotated on the diagonal axis between the two corners that were held. He drew on the cracker another arrow that also pointed north.

Next, he held the cracker on his right in similar fashion, with his right hand, and rotated it by pushing with his left forefinger on corner B. This time, however, instead of drawing an arrow that pointed north, he drew one that pointed south.

"Now we're all set," he said, smiling, "for an amusing stunt involving the symmetry rotations of a square. You'll note that on the left I have a cracker with a north arrow on both sides." He picked up the cracker with his left hand and rotated it several times to show that on both sides the arrow pointed north. "And on my right we have north and south arrows." He picked up the cracker with his right hand and rotated it rapidly several times to show that the two arrows pointed in opposite directions.

Victor returned the cracker to the table. Then, slowly and without altering their orientation, he switched the positions of the two crackers. "Please rotate them yourself," he requested. "I want you to verify the fact that the cracker with the two north arrows is now on my right, and the other cracker on my left."

He handed me each cracker and I rotated it in exactly the same way he had done, one in my right hand and one in my left. Yes, the crackers had been exchanged.

Victor placed the crackers in front of him, then snapped his fingers and commanded the crackers to return invisibly to their former positions. He rotated the cracker on his left. I was startled to see that the arrows now pointed north on both sides! And when he rotated the other cracker, its arrows jumped back and forth from north to south!

"Try it," Victor said. "You'll find that it works automatically. Actually, both crackers are exactly alike. The difference in appearance depends entirely on which hand is holding them. When you ask your spectator to check on the crackers, be sure he takes the cracker on your right in his left hand, and the cracker on your

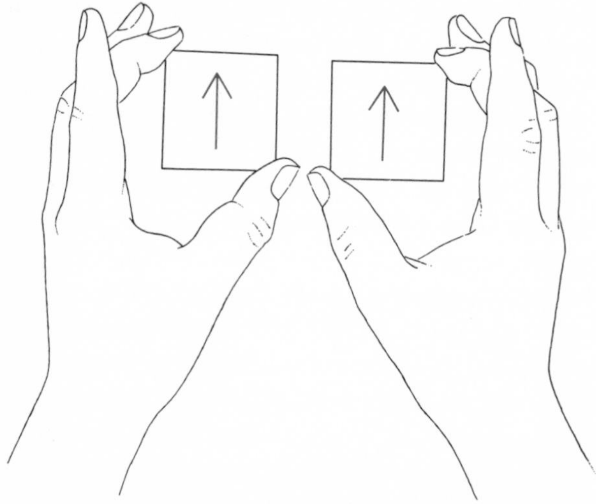

FIG. 42
How soda crackers are held for the trick of the transposed arrows.

left in his right hand. And see that he puts down the north-south cracker so the arrow on the top side points north."

I drained my glass. There was just enough left in the bottle for one more highball. The kitchen wobbled slightly.

"Now let *me* show *you* one," I said, taking another cracker from the box. "It's a test of probability. I'll toss this cracker into the air. If it falls rough side up, you get the rest of the bourbon. If it falls smooth side up, you get the rest of the bourbon. If it falls with neither side up" (I held the cracker perpendicular to the table but made no comment about it), "then *I* get the last drink."

Victor looked wary. "Okay," he said.

I squeezed the cracker in my fist and tossed the crumbs into the air.

Dead silence. Even the refrigerator stopped humming. "I observe that most of both sides came down on your head," Victor said at last, unsmiling. "And I must say it's a pretty crumby trick to play on an old friend."

ADDENDUM

THE GILBREATH principle and its use in the trick described were first explained by Norman Gilbreath in an article, "Magnetic Colors," in a magic periodical called *The Linking Ring*, Vol. 38, No. 5, page 60, July 1958. Since then, dozens of clever card tricks have been based on the simple principle. For those with access to magic journals, here are a few references:

Linking Ring, Vol. 38, No. 11, pages 54–58, January 1958. Tricks by Charles Hudson and Ed Marlo.

Linking Ring, Vol. 39, No. 3, pages 65–71, May 1959. Tricks by Charles Hudson, George Lord and Ron Edwards.

Ibidem (A Canadian magic periodical), No. 16, March 1959. Trick by Tom Ransom.

Ibidem, No. 26, September 1962. Trick by Tom Ransom.

Ibidem, No. 31, December 1965. Trick by Allan Slaight.

The principle can be proved informally as follows. When the deck is cut for a riffle shuffle, there are two possible situations: The bottom cards of the two halves are either the same color or different. Assume they are different. After the first card falls, the bottom cards of the two halves will then be the *same* color, and opposite to that of the card that fell. It makes no difference, therefore, whether the next card slips past the left or right thumb; in either case, a card of opposite color must fall on the previous one. This places on the table a pair of cards that do not match. The situation is now exactly as before. The bottom cards of the halves in the hands do not match. Whichever card falls, the remaining bottom cards will both have the opposite color. And so on. The argument repeats for each pair until the deck is exhausted.

Now suppose that the deck is initially cut so that the two bottom cards are the *same* color. Either card may fall first. The previous argument now applies to all the pairs of cards that follow. One last card will remain. It must, of course, be opposite in color to the first card that fell. When the deck is cut between two cards of the same color (that is, between the ordered pairs), the top and bottom cards of the deck are brought together, and all pairs are now intact.

There are many different ways of presenting the trick with the cards and the glass. Ron Edwards, of Rochester, New York, writes that he has nine cards selected at random and formed into a square. The spectator then places a miniature skull on one of the cards. There is a hole in the top of the skull into which Edwards places a rolled slip of paper on which he has written his prediction: the name of the center card. The proper instruction card is then taken from his pocket (the two cards are in different pockets). The instructions designate the positions (rather than names) of the cards to be removed at each step.

After this trick appeared in *Scientific American,* Hal Newton, Rochester, New York, worked out a version called "Voice from Another World" in which a phonograph record is played to give instructions to a spectator as he moves an object back and forth on nine cards that bear the names of the nine planets. The record can, of course, be played on either side. The trick was put on the market in 1962 by Gene Gordon's magic shop in Buffalo.

ANSWERS

THE CARD-SHUFFLED sentence deciphers as: "The smelling organs of fish have evolved in a great variety of forms." It is the first sentence of the last paragraph on page 73 of the article "The Homing Salmon," by Arthur D. Hasler and James A. Larsen, in *Scientific American* for August 1955.

CHAPTER TEN

□

The Four-Color Map Theorem

Hues
Are what mathematicians use
(While hungry patches gobble 'em)
For the 4-color problem.

A Clerihew by
J. A. Lindon, Surrey, England

OF ALL the great unproved conjectures of mathematics, the simplest — simple in the sense that a small child can understand it — is the famous four-color theorem of topology. How many colors are needed for coloring any map so that no two countries with a common border will have the same color? It is easy to construct maps that require four colors, and only a knowledge of elementary mathematics is required to follow a rigorous proof that five colors are sufficient. But are four colors both necessary and sufficient? To put it another way, is it possible to construct a map that will require five colors? Mathematicians who are interested in the matter think not, but they are not sure.

Every few months I receive in the mail a lengthy "proof" of the four-color theorem. In almost every case it turns out that the sender has confused the theorem with a much simpler one which states that it is impossible to draw a map of five regions in such a way that each region is adjacent to the other four. (Two regions that meet at a single point are not considered adjacent.) I myself contributed in a small way to this confusion by once writing a science-fiction story entitled "The Island of Five Colors," about an imaginary island divided by a Polish topologist into five regions that all had common borders. It is not difficult to prove that a map of this sort cannot be drawn. One might suppose that the four-color theorem for all maps would now follow automatically, but such is not the case.

To see why this is so, consider the simple map at *a* in Figure 43. (The actual shapes of the regions do not matter; only the manner in which they are connected is significant. The four-color theorem is topological precisely because it deals with a property of plane figures that is unaltered by distorting the surface on which they are inscribed.) What color shall we use for the blank region? Obviously we must color it either red or a fourth color. Suppose we take the second alternative and color it green, as shown at *b* in the illustration. Then we add another region. It is now impossible to complete the map without using a fifth color. Let us go back, then, to *a*, and instead of putting green for the blank region, use red. But this gets us into difficulty if two more regions touch the first four, as shown at *c*. Clearly fourth and fifth colors are necessary for the two blank areas. Does all this prove that five colors are necessary for some maps? Not at all. In both cases we can manage with four colors, but only by going back and altering the previous color scheme.

In coloring complicated maps, with dozens of regions, we find ourselves constantly running into blind alleys of this sort that require a retracing of steps. To prove the four-color theorem, therefore, one must show that in all cases such alterations can always be made successfully, or devise a procedure that will eliminate all such alterations in the process of coloring any map with four colors. Stephen Barr has suggested a delightful two-person topological game that is based on the difficulty of foreseeing these color cul-de-sacs. Player A draws a region. Player B colors it and adds a new region. Player A colors the new region and adds a third. This continues, with each player coloring the last region drawn by his opponent, until a player loses the game by being forced to use a fifth color. I know of no quicker way to recognize the difficulties involved in proving the four-color theorem than to engage in this curious game.

It is often said that cartographers were the first to realize that no more than four colors are required for any map, but this has been questioned by Kenneth O. May, a mathematician at Carleton College. After extensive research on the origin of the four-color theorem, May failed to find any statement of the theorem in early books on cartography, or any indication that the theorem was recognized. It seems to have been first formulated explicitly by Francis Guthrie, a student at Edinburgh. He mentioned it to

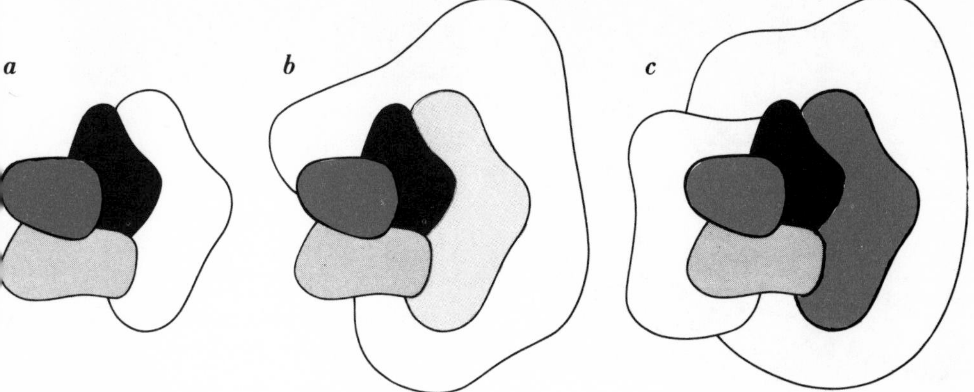

FIG. 43
In making a map with four colors it is often necessary to start over again with different colors.

his brother Frederick (who later became a chemist), and Frederick in turn passed it on, in 1852, to his mathematics teacher, Augustus de Morgan. The conjecture became well known after the great Arthur Cayley admitted in 1878 that he had worked on the theorem but had been unable to prove it.

In 1879 the British lawyer and mathematician Sir Alfred Kempe published what he believed to be a proof, and a year later he contributed to the British journal *Nature* an article with the overconfident title "How to Colour a Map with Four Colours." For ten years mathematicians thought the problem had been disposed of; then P. J. Heawood spotted a fatal flaw in Kempe's proof. Since that time the finest minds in mathematics have grappled unsuccessfully with the problem. The tantalizing thing about the theorem is that it *looks* as though it should be quite easy to prove. In his autobiographical book *Ex-Prodigy* Norbert Wiener writes that he has tried, like all mathematicians, to find a proof of the four-color theorem, only to find his proof crumble, as he expresses it, to fool's gold in his hands. As matters now stand, the theorem has been established for all maps with no more than 38 regions. This may seem like a small number, but it becomes less trivial when we realize that the number of topologically different maps with 38 or less regions would run to more than 10^{38}. Even a

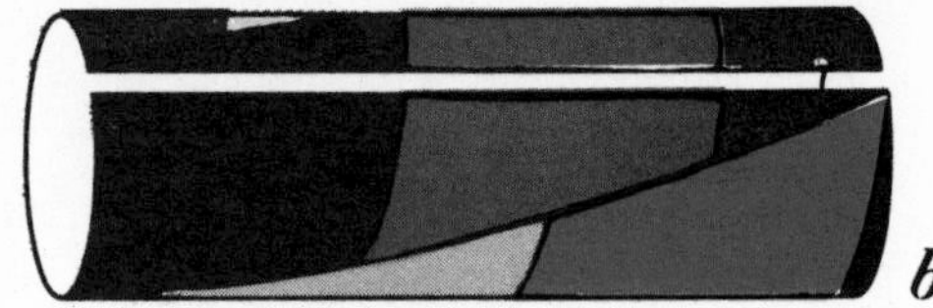

FIG. 44
Seven colors make a map on a torus (c). The sheet (a) is first rolled into a cylinder (b). The resulting torus is enlarged.

modern electronic computer would not be able to examine all these configurations in a reasonable length of time.

The lack of proof for the four-color theorem is made even more exasperating by the fact that analogous proofs have been found for surfaces much more complicated than the plane. (The surface of a sphere, by the way, is the same as a plane so far as this problem goes; any map on the sphere can be transformed to an equivalent plane map by puncturing the map inside any region and then flattening the surface.) On one-sided surfaces such as the Moebius strip, the Klein bottle and the projective plane it has been established that six colors are necessary and sufficient. On the surface of the torus, or anchor ring, the number is seven. Such a map is shown in Figure 44. Note that each region is bounded by six line segments and that every region is adjacent to the other six. In fact, the map-coloring problem has been solved for every higher surface that has been seriously investigated.

It is only when the theorem is applied to surfaces topologically equivalent to a plane or surface of a sphere that its proof continues to frustrate topologists; and what is worse still, there is no apparent reason why this should be so. There is something spooky about the way in which attempted proofs seem to be working out beautifully, only to develop an infuriating gap just as the deductive chain is about to be completed. No one can predict what the future will decide about this famous problem, but we can be sure that world fame awaits the first person who achieves one of three possible breakthroughs:

1. A map requiring five colors will be discovered. "If I be so bold as to make a conjecture," writes H. S. M. Coxeter in his

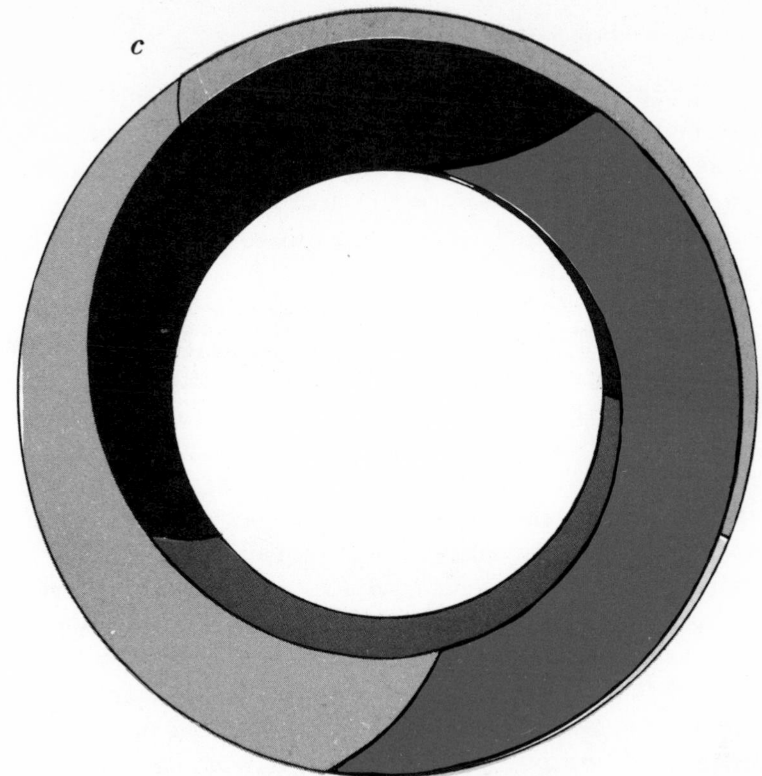

excellent article "The Four-Color Map Problem, 1840–1890," "I would guess that a map requiring five colors may be possible, but that the simplest such map has so many faces (maybe hundreds or thousands) that nobody, confronted with it, would have the patience to make all the necessary tests that would be required to exclude the possibility of coloring it with four colors."

2. A proof of the theorem will be found, possibly by a new technique that may suddenly unlock many another bolted door of mathematics.

3. The theorem will be proved impossible to prove. This may sound strange, but in 1931 Kurt Gödel established that in every deductive system complicated enough to include arithmetic there are mathematical theorems that are "undecidable" within the system. So far very few of the great unsolved conjectures of mathematics have been shown to be undecidable in this sense. Is the four-color theorem such a theorem? If so, it can be accepted as "true" only by adopting it, or some other undecidable theorem closely linked to it, as a new and unprovable postulate of an enlarged deductive system.

Unfortunately the proof that five colors are sufficient for plane maps, or that six or more colors are necessary and sufficient for certain higher surfaces, is too lengthy to include here. But perhaps the following clever proof of a two-color theorem will give the reader some notion of how one can go about establishing a map-coloring theorem.

Consider all possible maps on the plane that can be formed by straight lines. The ordinary checkerboard is a familiar example. A less regular pattern is shown in the left illustration of Figure 45. Are two colors sufficient for all such maps? The answer is yes, and it is easily shown. If we add another straight line (*e.g.*, the heavy black line in the same illustration) to any properly colored straight-line map, the line will divide the plane into two separate maps, each correctly colored when considered in isolation, but with pairs of like-color regions adjacent along the line. To restore a proper coloration to the entire map, all we have to do is exchange the two colors on one side (it doesn't matter which) of the line. This is shown in the right illustration. The map above the black line has been reversed, as though a negative print had been changed to a positive, and, as you can see, the new map is now properly colored.

To complete the proof, consider a plane that is divided into two regions by a single line. It can of course be constructed with two colors. We draw a second line and recolor the new map by reversing the colors on one side of the line. We draw a third line, and so on. Clearly this procedure will work for any number of lines, so by a method known as "mathematical induction" we have established a two-color theorem for all possible maps drawn

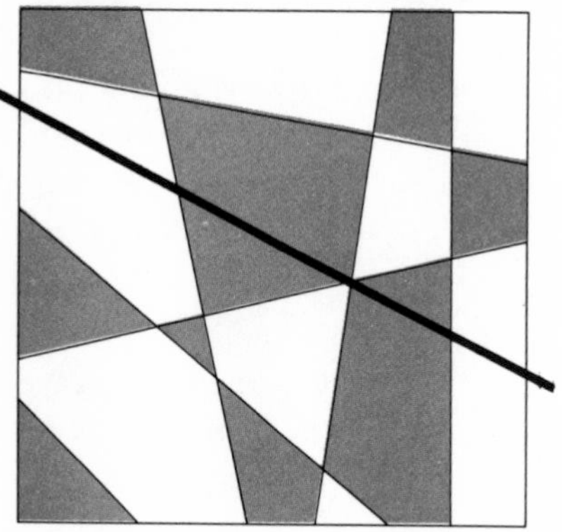

FIG. 45

Two colors suffice for any map drawn with lines that cut across the entire surface.

FIG. 46
Two colors also suffice for a map drawn with lines that either cut across or form closed curves.

with straight lines. The proof can be generalized to cover less rigid maps, such as the one in Figure 46, which are drawn with endless lines that either cut across the entire map or lie on it as simple closed curves. If we add a line that crosses the map, we reverse the colors on one side of the line as before. If the new line is a closed curve, we reverse the colors of all regions inside the curve or, if we prefer, the colors of regions outside the curve. The closed curves may also intersect themselves, but then the recoloring procedure becomes more complicated.

Note that all the two-color maps shown here have even vertices; that is, at each vertex an even number of lines meet. It can be proved that any map on the plane can be colored with two colors if and only if all its vertices are even. This is known as the "two-color map theorem." That it does not hold on the torus is easily seen by ruling a square sheet of paper into nine smaller squares (like a ticktacktoe field) and rolling it into a torus in the manner previously described. This checked doughnut has even vertices but requires three colors.

Now, more for amusement than for enlightenment, here are three map-coloring problems that are not difficult, although each has a "catch" element of some sort that makes the solution not quite what one would at first expect:

1. How many colors are required for the map in Figure 47 (devised by the English puzzlist Henry Ernest Dudeney) so that no two regions of the same color border on each other?

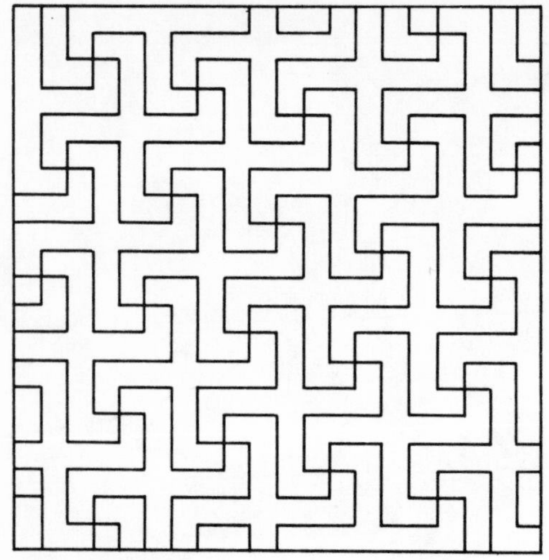

FIG. 47
How many colors are needed for this map?

2. Stephen Barr writes about the painter who wished to complete on a huge canvas the nonobjective work of art shown in outline in Figure 48. He decided to limit himself to four colors, and to fill each region with one solid color in such a way that there would be a different color on each side of every common border. Each region had an area of eight square feet, except for the top region, which was twice the size of the others. When he checked his paint supplies, he found that he had on hand only the following: enough red to cover 24 square feet, enough yellow to cover the same area, enough green to cover sixteen square feet and enough blue to paint eight square feet. How did he manage to complete his canvas?

3. Leo Moser, a mathematician at the University of Alberta, asks: How can a two-color map be drawn on a plane so that no matter where you place on it an equilateral triangle with a side of 1, all three vertices never lie on points of the same color?

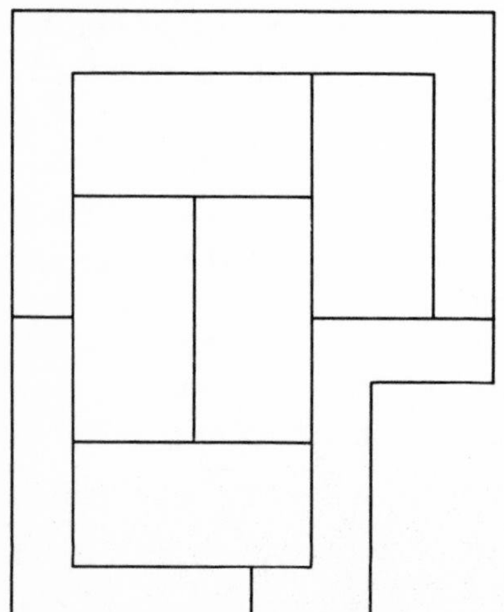

FIG. 48
How many are needed for this abstraction?

ADDENDUM

THE ASSERTION that five regions cannot be drawn on the plane so that every pair has a common border was made by Moebius in an 1840 lecture. He gave it in the form of a story about an Eastern prince who willed his kingdom to five sons on condition that it be divided into five regions, each bordering the others. The problem is equivalent to the following problem in graph theory: Is it possible to place five spots on the plane and join each to the others by straight lines that do not intersect? Proofs of impossibility are not difficult, and can be found in any book on elementary graph theory. An easy-to-follow proof is given by Heinrich Tietze in his chapter "On Neighboring Domains" in *Famous Problems of Mathematics*. Essentially the same proof is sketched by Henry Dudeney in his solution to problem 140 in *Mathematical Puzzles*. Dudeney goes on to argue, mistakenly, that this implies a proof of the four-color theorem.

The looseness of my language in speaking of the four-color theorem as "Gödel-undecidable" prompted the following letter from the British cosmologist Dennis Sciama (*Scientific American*, November 1960, page 21):

> *Sirs:*
>
> *I have been enjoying Martin Gardner's article on the four-color problem. Actually it is impossible to prove that it is impossible to prove the theorem. For if the theorem is false, this can undoubtedly be shown explicitly by exhibiting a map that cannot be colored with four colors. Hence if the theorem is unprovable it must be true. This means that we cannot prove it to be unprovable, for this is tantamount to proving it to be true, which is a contradiction.*
>
> *The same remark holds for any theorem whose falsity could be demonstrated by a gegenbeispiel;* e.g., *Fermat's last theorem. Such theorems may be unprovable, but only if they are true. We can then never know that they are unprovable, so that mathematicians would endlessly try to prove them. This is a terrifying state of affairs. Doing physics might seem to be a good alternative, but Gödelry may invade that realm yet. . . .*

The situation is a bit less terrifying when we realize that a theorem that is Gödel-undecidable within a given deductive system can always be decided metamathematically by enlarging the system. If the four-color theorem is ever shown to be Gödel-undecidable within a system resting on certain postulates of topology and set theory, it automatically becomes "true" (as Sciama makes clear), but "true" in the metamathematical sense of being decidable in a larger system, perhaps a system in which the map theorem itself is a new postulate.

ANSWERS

THE ANSWERS to the three map-coloring problems follow (the first two answers refer to illustrations that accompanied the problems):

1. The swastika map could be colored with two colors were it not for one small line at the lower left corner. At this spot three regions touch one another, so three colors are required.

2. The artist colored his abstraction by mixing all his blue paint with one-third of his red paint to obtain enough purple to color sixteen square feet of canvas. After the large region at the top of the canvas and the area in the center are painted yellow, it is a simple matter to color the remaining regions red, green and purple.

3. To color the plane with two colors so that no three points of the same color mark the corners of an equilateral triangle with a side of 1, the simplest method is to divide the plane into parallel stripes, each with a width of $\sqrt{3}/2$, then color them alternately black and white as shown in Figure 49. This does not solve the problem, however, until the concept of open and closed sets is introduced. A continuum of real numbers — say from 0 to 1 — is called a closed interval if it includes 0 and 1, and an open interval if it excludes them. If it includes one and not the other, it is said to be closed at one end and open at the other.

The stripes on the map are closed along their left edge; open along their right. The black stripe on the left has a width that starts at 0, measured on the line below the map, and goes to $\sqrt{3}/2$. It includes 0, but does not include $\sqrt{3}/2$. The next stripe has a width that includes $\sqrt{3}/2$ but does not include $2\sqrt{3}/2$, and so on for the other stripes. In other words, each vertical line belongs only to the stripe on its right. This is necessary to take care of cases in which the triangle, shown in color, lies with all three of its corners on boundary lines.

Leo Moser, of the University of Alberta, who sent this problem, writes that it is not known how many colors are required for coloring the plane so that no *two* points, a unit distance apart, lie on the same color. Four colors have been shown necessary, and seven sufficient. (That seven are sufficient is evident from a regular tiling of hexagons, each with the radius of its circumscribing circle a trifle less than unity, and each surrounded by hexagons that differ in color from it and from each other.) The gap between four and seven is so large that the problem seems a long way from being solved.

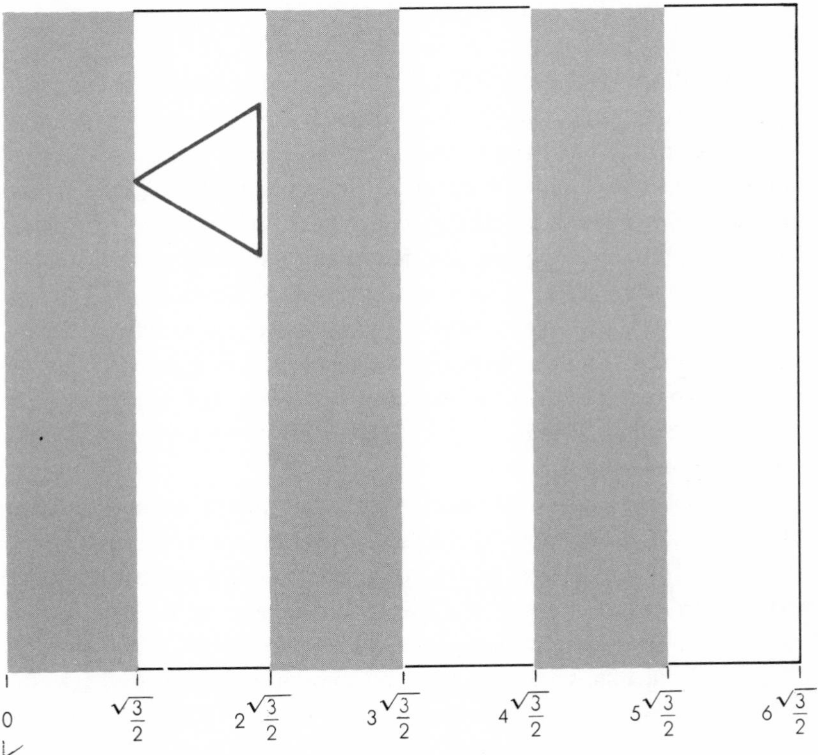

FIG. 49
Solution to the problem of the triangle and the two-color map.

CHAPTER ELEVEN

□

Mr. Apollinax Visits New York

When Mr. Apollinax visited the United States
His laughter tinkled among the teacups.

— T. S. Eliot

P. BERTRAND APOLLINAX, the brilliant protégé of the celebrated French mathematician Nicolas Bourbaki, was little known even in France until the spring of 1960. It was then, as everyone knows, that the mathematical world was shattered by the disclosure, in a French mathematical journal, of what is now known as the Apollinax function. By means of this remarkable function Apollinax was able at one stroke to (1) prove Fermat's last theorem, (2) provide a counterexample (a map with 5,693 regions) to the famous four-color theorem of topology, (3) lay the groundwork for Channing Cheetah's discovery, three months later, of a 5,693-digit integer — the first of its kind known — that is both perfect and odd.

The reader will understand my excitement when Professor Cheetah, of New York University, invited me to his apartment for an afternoon tea at which Apollinax would be guest of honor. (Cheetah's apartment is in Greenwich Village, in a large brownstone building off Fifth Avenue. The building is owned by Mrs. Orville Phlaccus, widow of the well-known financier, and is called Phlaccus Palace by students at nearby N.Y.U.) When I arrived, the tea was in full swing. I recognized several members of the N.Y.U. mathematics faculty and guessed that most of the younger people present were graduate students.

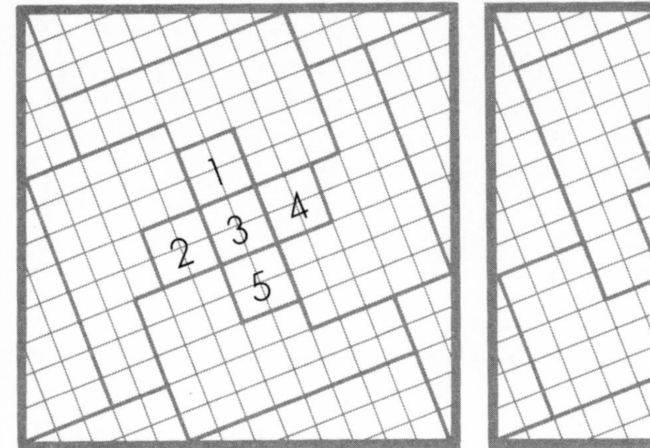

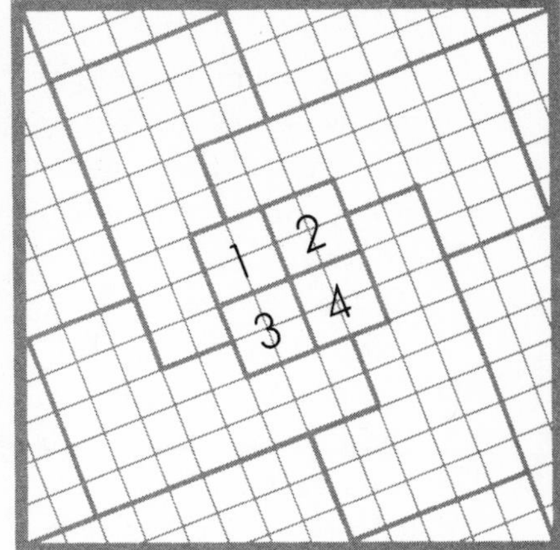

FIG. 50
The mystery of the disappearing tile.

There was no mistaking Apollinax. He was the obvious center of attention: a bachelor in his early thirties, tall, with rugged features that could not be called handsome but nevertheless conveyed a strong impression of physical virility combined with massive intellect. He had a small black goatee and rather large ears with prominent Darwin points. Under his tweedy jacket he sported a bright red vest.

While Mrs. Phlaccus served me a cup of tea, I heard a young woman say: "That silver ring on your finger, Mr. Apollinax. Isn't it a Moebius strip?"

He removed the ring and handed it to her. "Yes. It was made by an artist friend of mine who has a jewelry shop on the Left Bank in Paris." He spoke with a husky French accent.

"It's crazy," the girl said as she handed back the ring. "Aren't you afraid it will twist around and your finger will disappear?"

Apollinax chuckled explosively. "If you think that's crazy, then I have something here you'll think even crazier." He reached into his side pocket and took out a square, flat wooden box. It was filled with seventeen white plastic tiles that fitted snugly together [*see Fig. 50, left*]. The tiles were of such thickness that the five small pieces in the center were cubes. Apollinax called attention to the number of cubes, dumped the tiles onto a nearby table, then quickly replaced them in the box in the manner shown in the illustration at right. They fitted snugly as before. But now there were only four cubes. One cube had completely vanished!

The young woman stared at the pattern with disbelief, then at Apollinax, who was shaking with high-pitched laughter. "May I study this for a while?" she asked, taking the box from his hand. She carried it off to a quiet corner of the room.

"Who's the chick?" Apollinax said to Cheetah.

"I beg your pardon?" replied the professor.

"The girl in the sweat shirt."

"Oh. Her name is Nancy Ellicott. A Boston girl. She's one of our undergraduate math majors."

"Very attractive."

"You think so? I've never seen her wear anything but dungarees and that same dirty sweat shirt."

"I like your Village nonconformists," Apollinax said. "They're all so much alike."

"Sometimes," remarked someone in the group, "it's hard to distinguish nonconformity from neurosis."

"That reminds me," I said, "of a mathematical riddle I just heard. What's the difference between a psychotic and a neurotic?"

Nobody said anything.

"A psychotic," I went on, "thinks that two plus two is five. A neurotic knows that it's four, but it makes him nervous."

There was some polite laughter, but Apollinax looked grave. "He has good reason to be nervous. Wasn't it Alexander Pope who wrote: 'Ah why, ye gods! should two and two make four?' Why indeed? Who can say why tautologies are tautological? And who can say that even simple arithmetic is free from contradiction?" He took a small notebook from his pocket and jotted down the following infinite series:

$$4 - 4 + 4 - 4 + 4 - 4 + 4 \ldots$$

"What," he asked, "is the sum of this series? If we group the numbers like this,

$$(4 - 4) + (4 - 4) + (4 - 4) \ldots$$

the sum is obviously zero. But if we group them so,

$$4 - (4 - 4) - (4 - 4) - (4 - 4) \ldots$$

the sum is clearly four. Suppose we try them still another way:

$$4 - (4 - 4 + 4 - 4 + 4 - 4 \ldots)$$

Now the sum of the series is four minus the sum of the same series. In other words, twice the sum is equal to four, so the sum must be equal to half of four, or two."

I started to make a comment, but Nancy pushed her way back through the group and said: "These tiles are driving me batty. What happened to that fifth cube?"

Apollinax laughed until his eyes teared. "I'll give you a hint, my dear. Perhaps it slid off into a higher dimension."

"Are you pulling my leg?"

"I wish I were," he sighed. "The fourth dimension, as you know, is an extension along a fourth coordinate perpendicular to the three coordinates of three-dimensional space. Now consider a cube. It has four main diagonals, each running from one corner through the cube's center to the opposite corner. Because of the cube's symmetry, each diagonal is clearly at right angles to the other three. So why shouldn't a cube, if it feels like it, slide along a fourth coordinate?"

"But my physics teacher," Nancy said with a frown, "told us that *time* was the fourth dimension."

"Nonsense!" Apollinax snorted. "General relativity is as dead as the dodo. Hasn't your professor heard about Hilbert Dongle's recent discovery of a fatal flaw in Einstein's theory?"

"I doubt it," Nancy replied.

"It's easy to explain. If you spin a sphere of soft rubber rapidly, what happens to its equator? It bulges. In relativity theory, you can explain the bulge in two different ways. You can assume that the cosmos is a fixed frame of reference — a so-called inertial system. Then you say that the sphere rotates and inertia makes the equator bulge. Or you can make the sphere a fixed frame of reference and regard the entire cosmos as rotating. Then you say that the masses of the moving stars set up a gravitational tensor field that exerts its strongest pull on the equator of the motionless ball. Of course—"

"I would put it a bit differently," Cheetah interrupted. "I would say that there is a relative movement of sphere and stars, and this relative motion causes a certain change in the space-time structure of the universe. It is the pressure, so to speak, of this space-time matrix that produces the bulge. The bulge can be viewed either as a gravitational or inertial effect. In both cases the field equations are exactly the same."

"Very good," Apollinax replied. "Of course, this is exactly what Einstein called the principle of equivalence — the equivalence of gravity and inertia. As Hans Reichenbach liked to put it, there's no truth distinction between the two. But now let me ask you this: Does not relativity theory make it impossible for physical bodies to have relative motions greater than the speed of light? Yet if we make the rubber ball our fixed frame of reference, it takes only a slow spin of the ball to give the moon a relative motion much faster than the speed of light."

Cheetah did a slow double-take.

"You see," Apollinax continued, "we just can't keep the sphere still while we spin the universe around it. This means that we have to regard the ball's spin as absolute, not relative. Astronomers run into the same sort of difficulty with what they call the transverse Doppler effect. If the earth rotates, the relative transverse velocity between the observatory and a ray of light from a distant star is very small, so the Doppler shift is small. But if you view the cosmos as rotating, then the transverse velocity of the distant star relative to the observatory is very great, and the Doppler shift would have to increase accordingly. Since the transverse Doppler shift is small, we must assume it is the earth that rotates. Of course, this defenestrates relativity theory."

"Then," Cheetah mumbled, looking a trifle pale, "how do you account for the fact that the Michelson-Morley experiment failed to detect any motion of the earth relative to a fixed space?"

"Quite simple," Apollinax said. "The universe is infinite. The earth spins around the sun, the sun speeds through the galaxy, the galaxy gallumphs along relative to other galaxies, the galaxies are in galactic clusters that move relative to other clusters, and the clusters are parts of superclusters. The hierarchy is endless. Add together an infinite series of vectors, of random speeds and directions, and what happens? They cancel each other out. Zero and infinity are close cousins. Let me illustrate."

He pointed to a large vase on the table. "Imagine that vase

empty. We start filling it with numbers. If you like, you can think of small counters with numbers on them. At one minute to noon we put the numbers 1 through 10 into the vase, then take out number 1. At one-half minute to noon, we put in numbers 11 to 20 and take out number 2. At one-third minute to noon we put in 21 to 30, take out 3. At one-fourth minute to noon we put in 31 to 40, take out 4. And so on. How many numbers are in the vase at noon?"

"An infinity," said Nancy. "Each time you take one out, you put in ten."

Apollinax cackled like an irresponsible hen. "There would be *nothing* in the vase! Is 4 in the vase? No, we took it out on the fourth operation. Is 518 in the vase? No, it came out on the 518th operation. The numbers in the vase at noon form an empty set. You see how close infinity is to zero?"

Mrs. Cheetah approached us, bearing a tray with assorted cookies and macaroons. "I think I shall exercise Zermelo's axiom of choice," said Apollinax, tugging on his goatee, "and take one of each kind."

"If you think relativity theory is dead," I said a few minutes later, "what is your attitude toward modern quantum theory? Do you think there's a fundamental randomness in the behavior of the elementary particles? Or is the randomness just an expression of our ignorance of underlying laws?"

"I accept the modern approach," he said. "In fact, I go much further. I agree with Karl Popper that there are *logical* reasons why determinism can no longer be taken seriously."

"That's hard to believe," someone said.

"Well, let me put it this way. There are portions of the future that *in principle* can never be predicted correctly, even if one possessed total information about the state of the universe. Let me demonstrate."

He took a blank file card from his pocket, then, holding it so no one could see what he was writing, he scribbled something on the card and handed it to me, writing side down. "Put that in your right trouser pocket."

I did as he directed.

"On that card," he said, "I've described a future event. It hasn't taken place yet, but it positively either will or will not take place before" — he glanced at his wrist watch — "before six o'clock."

He took another blank card from his pocket and handed it to me. "I want you to try to guess whether the event I just described will take place. If you think it will, write 'Yes' on the card you hold. If you think it won't, write 'No.' "

I started to write, but Apollinax caught my wrist. "Not yet, old chap. If I see your prediction, I might do something to make it fail. Wait until my back is turned, and don't let anyone see what you write." He spun around and looked at the ceiling until I had finished writing. "Now put the card in your left pocket, where no one can see it."

He turned to face me again. "I don't know your prediction. You don't know what the event is. Your chance of being right is one in two."

I nodded.

"Then I'll make you the following bet. If your prediction is wrong, you must give me ten cents. If it's right, I'll give you one million dollars."

Everyone looked startled. "It's a deal," I said.

"While we're waiting," Apollinax said to Nancy, "let's go back to relativity theory. Would you care to know how you can always wear a relatively clean sweat shirt, even if you own only two sweat shirts and never wash either of them?"

"I'm all ears," she said, smiling.

"You have other features," he said, "and very pretty ones too. But let me explain about those sweat shirts. Wear the cleanest one, say sweat shirt A, until it becomes dirtier than B. Then take it off and put on the relatively clean sweat shirt B. The instant B is dirtier than A, take off B and put on the relatively clean sweat shirt A. And so on."

Nancy made a face.

"I really can't wait here until six," Apollinax said. "Not on a warm spring evening in Manhattan. Would you by any chance know if Thelonious Monk is playing anywhere in the city tonight?"

Nancy's eyes opened wide. "Why, yes, he's playing right here in the Village. Do you like his style?"

"I dig it," Apollinax said. "And now, if you'll kindly direct me to a nearby restaurant, where I shall pay for your dinner, we will eat, I will explain the mystery of the tiles, then we will go listen to the Monk."

After Apollinax had left, with Nancy on his arm, word of the prediction bet spread rapidly around the room. When six o'clock arrived, everyone gathered around to see what Apollinax and I had written. He was right. The event was logically unpredictable. I owed him a dime.

The reader may enjoy trying to figure out just what future event Apollinax described on that card.

ADDENDUM

MANY READERS took Apollinax seriously (even though I said he was a protégé of Bourbaki, the well-known, nonexistent French mathematician) and wrote to ask where they could find out about the "Apollinax function." Both Apollinax and Nancy, as well as others at the tea, are straight out of T. S. Eliot's two poems, "Mr. Apollinax" and "Nancy," which appear on facing pages in Eliot's *Collected Poems: 1909–1962* (Harcourt Brace, 1963).

"Mr. Apollinax," by the way, is a poem about Bertrand Russell. When Russell visited Harvard in 1914, Eliot attended his lectures on logic, and the two met at a tea; the tea Eliot describes in his poem. A mathematician at Trinity College, Cambridge, wrote to ask me if the name "Phlaccus" was a portmanteau word combining "flaccid" and "phallus"; I mention this as a minor contribution to Eliot exegesis. Hilbert Dongle derives from Herbert Dingle, the British physicist who has been arguing in recent years that if the clock paradox of relativity theory is true, then relativity isn't. (See my chapter on the clock paradox in *Relativity for the Million,* now a Pocket Books paperback.) Thelonious Monk is Thelonious Monk.

Apollinax's reasoning about Nancy's dirty sweat shirt is borrowed from a small poem by Piet Hein, who is mentioned earlier in the chapter on braids. The paradox about the numbers in the vase comes from J. E. Littlewood's *A Mathematician's Miscellany.* It illustrates a case in which the subtraction of the transfinite number aleph-null from ten times aleph-null results in zero. If the numbered counters are taken out of the vase in the order 2, 4, 6, 8 . . . , an aleph-null infinity remains, namely, all the odd numbers. One can also remove an infinite set of counters in such a way as to leave any desired finite number of counters. If one wishes to leave, say, exactly three counters, he merely takes out numbers in serial order, but beginning with 4. The situation is an amusing illustration of the fact that when aleph-null is taken from aleph-null, the result is indeterminate; it can be made zero, infinity, or any desired positive integer, depending on the nature of the two infinite sets that are involved.

The pattern for the vanishing-cube paradox is one that I based on a little-known principle discovered by Paul Curry, of New York City, and which is discussed at length in the chapters on "Geometrical Vanishes" in my Dover paperback, *Mathematics, Magic and Mystery.*

My dramatization of the prediction paradox as a bar bet was first published in *Ibidem,* a Canadian magic magazine, No. 23, March 1961, page 23. I contributed a slightly different version, involving a card mailed to a friend, to *The British Journal for the Philosophy of Science,* Vol. 13, page 51, May 1962.

ANSWERS

THE PARADOX of the tiles, demonstrated by P. Bertrand Apollinax, is explained as follows. When all seventeen tiles are formed into a square, the sides of the square are not absolutely straight but convex by an imperceptible amount. When one cube is removed and the sixteen tiles re-formed into a square, the sides of the square are concave by the same imperceptible amount. This accounts for the apparent change in area. To dramatize the paradox, Apollinax performed a bit of sleight of hand by palming the fifth cube as he rearranged the pattern of the tiles.

In his prediction bet the event that Apollinax described on the file card was: "You will place in your left trouser pocket a card on which you have written the word 'No.' " The simplest presentation of the same paradox is to ask someone to predict, by saying yes or no, whether the next word that he utters will be no. Karl R. Popper's reasons for thinking that part of the future is in principle unpredictable are not based on this paradox, which is simply a version of the old liar paradox, but on much deeper considerations. These considerations are given in Popper's "Indeterminism in Quantum Physics and in Classical Physics," in *The British Journal for the Philosophy of Science,* Vol. 1, No. 2 and 3, 1950, and will be discussed more fully in his forthcoming book *Postscript: After Twenty Years.* A prediction paradox essentially the same as Apollinax's, except that it involves a computer and electric fan instead of a person and card, is discussed in Chapter 11 of John G. Kemeny's *A Philosopher Looks at Science,* published by D. Van Nostrand in 1959.

The paradox of the infinite series of fours, alternately added and subtracted, is explained by the fact that the sum of this series does not converge but oscillates back and forth between the values of zero and four. To explain the rotation paradoxes would require too deep a plunge into relativity theory. For a stimulating presentation of a modern approach to these classic difficulties, Dennis Sciama's recent book, *The Unity of the Universe,* published by Doubleday & Company, Inc., is recommended.

CHAPTER TWELVE

□

Nine Problems

1. THE GAME OF HIP

THE GAME of "Hip," so named because of the hipster's reputed disdain for "squares," is played on a six-by-six checkerboard as follows:

One player holds eighteen red counters; his opponent holds eighteen black counters. They take turns placing a single counter on any vacant cell of the board. Each tries to avoid placing his counters so that four of them mark the corners of a square. The square may be any size and tipped at any angle. There are 105 possible squares, a few of which are shown in Figure 51.

A player wins when his opponent becomes a "square" by forming one of the 105 squares. The game can be played on a board with actual counters, or with pencil and paper. Simply

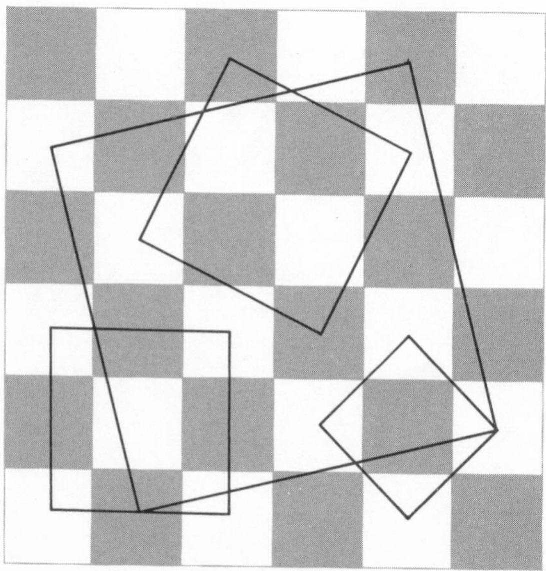

FIG. 51 Four of the 105 ways to become "square" in the game of Hip.

draw the board, then register moves by marking X's and O's on the cells.

For months after I had devised this game I believed that it was impossible for a draw to occur in it. Then C. M. McLaury, a mathematics student at the University of Oklahoma, demonstrated that the game could end in a draw. The problem is to show how the game can be drawn by dividing the 36 cells into two sets of eighteen each so that no four cells of the same set mark the corners of a square.

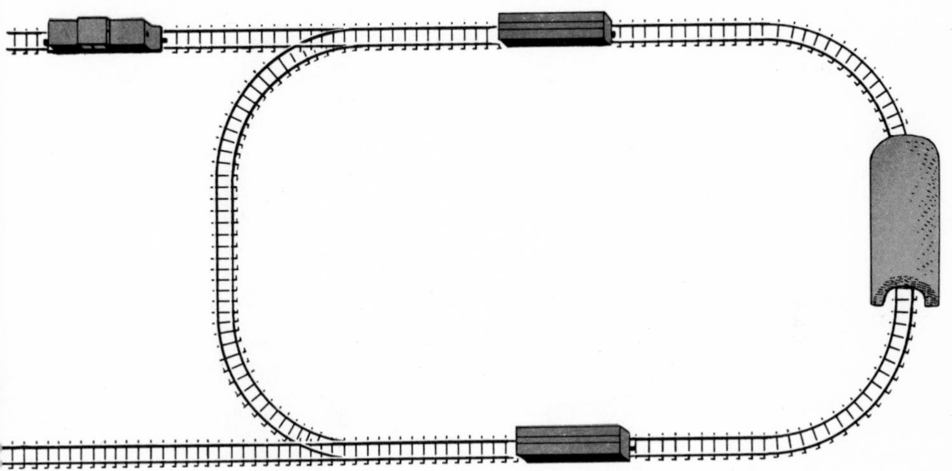

FIG. 52
A puzzle in operations research.

2. A SWITCHING PUZZLE

THE EFFICIENT switching of railroad cars often poses frustrating problems in the field of operations research. The switching puzzle depicted in Figure 52 is one that has the merit of combining simplicity with surprising difficulty.

The tunnel is wide enough to accommodate the locomotive but not wide enough for either car. The problem is to use the locomotive for switching the positions of cars A and B, then return the locomotive to its original spot. Each end of the locomotive can be used for pushing or pulling, and the two cars may, if desired, be coupled to each other.

The best solution is the one requiring the fewest operations. An "operation" is here defined as any movement of the locomotive between stops, assuming that it stops when it reverses direction, meets a car to push it or unhooks from a car it has been pulling. Movements of the two switches are not counted as operations.

A convenient way to work on the puzzle is to place a penny, a dime and a nickel on the illustration and slide them along the tracks, remembering that only the coin representing the locomotive can pass through the tunnel. In the illustration, the cars were drawn in positions too close to the switches. While working on the problem, assume that both cars are far enough east along the track so that there is ample space between each car and switch to accommodate both the locomotive and the other car.

No "flying switch" maneuvers are permitted. For example, you are not permitted to turn the switch quickly just after the engine has pushed an unattached car past it, so that the car goes one way and the engine, without stopping, goes another way.

3. BEER SIGNS ON THE HIGHWAY

SMITH DROVE at a steady clip along the highway, his wife beside him. "Have you noticed," he said, "that those annoying signs for Flatz beer seem to be regularly spaced along the road? I wonder how far apart they are."

Mrs. Smith glanced at her wrist watch, then counted the number of Flatz beer signs they passed in one minute.

"What an odd coincidence!" exclaimed Smith. "When you multiply that number by ten, it exactly equals the speed of our car in miles per hour."

Assuming that the car's speed is constant, that the signs are equally spaced and that Mrs. Smith's minute began and ended with the car midway between two signs, how far is it between one sign and the next?

4. THE SLICED CUBE AND THE SLICED DOUGHNUT

AN ENGINEER, noted for his ability to visualize three-dimensional structure, was having coffee and doughnuts. Before he dropped a sugar cube into his cup, he placed the cube on the table and thought: If I pass a horizontal plane through the cube's center, the cross section will of course be a square. If I pass it vertically through the center and four corners of the cube, the cross section will be an oblong rectangle. Now suppose I cut the cube this way with the plane. . . . To his surprise, his mental image of the cross section was a regular hexagon.

How was the slice made? If the cube's side is half an inch, what is the side of the hexagon?

After dropping the cube into his coffee, the engineer turned his attention to a doughnut lying flat on a plate. "If I pass a plane horizontally through the center," he said to himself, "the cross section will be two concentric circles. If I pass the plane vertically through the center, the section will be two circles separated by the width of the hole. But if I turn the plane so. . . ." He whistled with astonishment. The section consisted of two perfect circles that intersected!

How was this slice made? If the doughnut is a perfect torus, three inches in outside diameter and with a hole one inch across, what are the diameters of the intersecting circles?

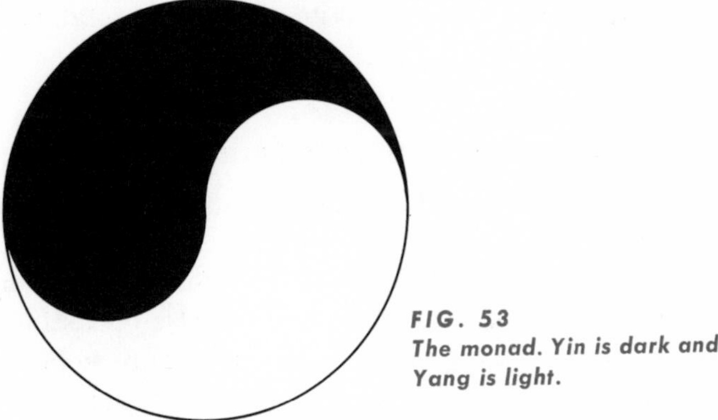

FIG. 53
The monad. Yin is dark and Yang is light.

5. BISECTING YIN AND YANG

TWO MATHEMATICIANS were dining at the Ying and Yang, a Chinese restaurant on West Third Street in Manhattan. They chatted about the symbol on the restaurant's menu [*see Fig. 53*].

"I suppose it's one of the world's oldest religious symbols," one of them said. "It would be hard to find a more attractive way to symbolize the great polarities of nature: good and evil, male and female, inflation and deflation, integration and differentiation."

"Isn't it also the symbol of the Northern Pacific Railway?"

"Yes. I understand that one of the chief engineers of the railroad saw the emblem on a Korean flag at the Chicago World's Fair in 1893 and urged his company to adopt it. He said it symbolized the extremes of fire and water that drove the steam engine."

"Do you suppose it inspired the construction of the modern baseball?"

"I wouldn't be surprised. By the way, did you know that there is an elegant method of drawing one straight line across the circle so that it exactly bisects the areas of the Yin and Yang?"

Assuming that the Yin and Yang are separated by two semicircles, show how each can be simultaneously bisected by the same straight line.

6. THE BLUE-EYED SISTERS

IF YOU HAPPEN to meet two of the Jones sisters (this assumes that the two are random selections from the set of all the Jones sisters), it is an exactly even-money bet that both girls will be blue-eyed. What is your best guess as to the total number of blue-eyed Jones sisters?

7. HOW OLD IS THE ROSE-RED CITY?

TWO PROFESSORS, one of English and one of mathematics, were having drinks in the faculty club bar.

"It is curious," said the English professor, "how some poets can write one immortal line and nothing else of lasting value. John William Burgon, for example. His poems are so mediocre that no one reads them now, yet he wrote one of the most marvelous lines in English poetry: 'A rose-red city half as old as Time.' "

The mathematician, who liked to annoy his friends with improvised brainteasers, thought for a moment or two, then raised his glass and recited:

"A rose-red city half as old as Time.
One billion years ago the city's age
Was just two-fifths of what Time's age will be
A billion years from now. Can you compute
How old the crimson city is today?"

The English professor had long ago forgotten his algebra, so he quickly shifted the conversation to another topic, but readers of this department should have no difficulty with the problem.

8. TRICKY TRACK

THREE HIGH SCHOOLS — Washington, Lincoln and Roosevelt — competed in a track meet. Each school entered one man, and one only, in each event. Susan, a student at Lincoln High, sat in the bleachers to cheer her boy friend, the school's shot-put champion.

When Susan returned home later in the day, her father asked how her school had done.

"We won the shot-put all right," she said, "but Washington High won the track meet. They had a final score of 22. We finished with 9. So did Roosevelt High."

"How were the events scored?" her father asked.

"I don't remember exactly," Susan replied, "but there was a certain number of points for the winner of each event, a smaller number for second place and a still smaller number for third place. The numbers were the same for all events." (By "number" Susan of course meant a positive integer.)

"How many events were there altogether?"

"Gosh, I don't know, Dad. All I watched was the shot-put."

"Was there a high jump?" asked Susan's brother.

Susan nodded.

"Who won it?"

Susan didn't know.

Incredible as it may seem, this last question can be answered with only the information given. Which school won the high jump?

9. TERMITE AND 27 CUBES

IMAGINE a large cube formed by gluing together 27 smaller wooden cubes of uniform size [*see Fig. 54*]. A termite starts at the center of the face of any one of the outside cubes and bores a path that takes him once through every cube. His movement is always parallel to a side of the large cube, never diagonal.

Is it possible for the termite to bore through each of the 26 outside cubes once and only once, then finish his trip by entering the central cube for the first time? If possible, show how it can be done; if impossible, prove it.

FIG. 54
The problem of the termite and the cube.

It is assumed that the termite, once it has bored into a small cube, follows a path entirely within the large cube. Otherwise, it could crawl out on the surface of the large cube and move along the surface to a new spot of entry. If this were permitted, there would, of course, be no problem.

ANSWERS

1. Figure 55 shows the finish of a drawn game of Hip. This beautiful, hard-to-find solution was first discovered by C. M. McLaury, a mathematics student at the University of Oklahoma to whom I had communicated the problem by way of Richard Andree, one of his professors.

Two readers (William R. Jordan, Scotia, New York, and Donald L. Vanderpool, Towanda, Pennsylvania) were able to show, by an exhaustive enumeration of possibilities, that the solution is unique except for slight variations in the four border cells indicated by arrows. Each cell may be either color, provided all four are not the same color, but since each player is limited in the game to eighteen pieces, two of these cells must be one color, two the other color. They are arranged here so that no matter how the square is turned, the pattern is the same when inverted.

The order-6 board is the largest on which a draw is possible. This was proved in 1960 by Robert I. Jewett, then a graduate student at the University of Oregon. He was able to show that a draw is impossible on the order-7, and since all higher squares contain a seven-by-seven subsquare, draws are clearly impossible on them also.

As a playable game, Hip on an order-6 board is strictly for the squares. David H. Templeton, professor of chemistry at the University of California's Lawrence Radiation Laboratory in Berkeley, pointed out that the second player can always force a draw by playing a simple symmetry strategy. He can either make each move so that it matches his opponent's last move by reflection across a parallel bisector of the board, or by a 90-degree rotation about the board's center. (The latter strategy could lead to the draw depicted.) An alternate strategy is to play in the corresponding opposite cell on a line from the opponent's last move and across the center of the board. Second-player draw strategies were also sent by Allan W. Dickinson, Richmond Heights, Missouri, and Michael Merritt, a student at Texas A. & M. College. These strategies apply to all even-order fields, and since no draws are possible on such fields higher than 6, the strategy guarantees a win for the second player on all even-order boards of 8 or higher. Even on the order-6, a reflection strategy across a parallel bisector is sure to win, because the unique draw pattern does not have that type of symmetry.

Symmetry play fails on odd-order fields because of the central cell. Since nothing is known about strategies on odd-order boards, the order-7 is the best field for actual play. It cannot end in a draw, and no one at present knows whether the first or second player wins if both sides play rationally.

In 1963 Walter W. Massie, a civil engineering student at Worcester Polytechnic Institute, devised a Hip-playing program for the IBM 1620 digital computer, and wrote a term paper about it. The program allows the computer to play first or second on any square field of orders 4 through 10. The computer takes a random cell if it moves first. On other plays, it follows a reflection strategy except when a reflected move forms a square, then it makes random choices until it finds a safe cell.

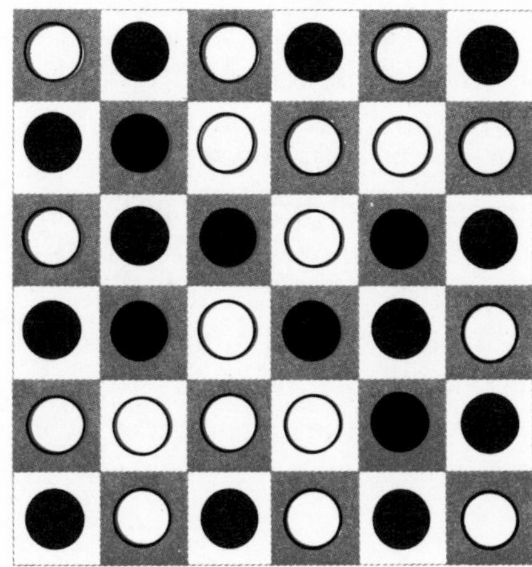

FIG. 55
Answer to the problem of the drawn game of Hip.

On all square fields of order n, the number of different squares that can be formed by four cells is $(n^4 - n^2)/12$. The derivation of this formula, as well as a formula for rectangular boards, is given in Harry Langman, *Play Mathematics*, Hafner, 1962, pages 36–37.

As far as I know, no studies have been made of comparable "triangle-free" colorings on triangular lattice fields.

2. The locomotive can switch the positions of cars A and B, and return to its former spot, in sixteen operations:

1. Locomotive moves right, hooks to car A.
2. Pulls A to bottom.
3. Pushes A to left, unhooks.
4. Moves right.
5. Makes a clockwise circle through tunnel.
6. Pushes B to left. All three are hooked.
7. Pulls A and B to right.
8. Pushes A and B to top. A is unhooked from B.
9. Pulls B to bottom.
10. Pushes B to left, unhooks.
11. Circles counterclockwise through tunnel.
12. Pushes A to bottom.
13. Moves left, hooks to B.
14. Pulls B to right.
15. Pushes B to top, unhooks.
16. Moves left to original position.

This procedure will do the job even when the locomotive is not permitted to pull with its front end, provided that at the start the locomotive is placed with its back toward the cars.

Howard Grossman, New York City, and Moises V. Gonzalez, Miami, Florida, each pointed out that if the lower siding is eliminated completely, the problem can still be solved, although two additional moves are required, making eighteen in all. Can the reader discover how it is done?

3. The curious thing about the problem of the Flatz beer signs is that it is not necessary to know the car's speed to determine the spacing of the signs. Let x be the number of signs passed in one minute. In an hour the car will pass $60x$ signs. The speed of the

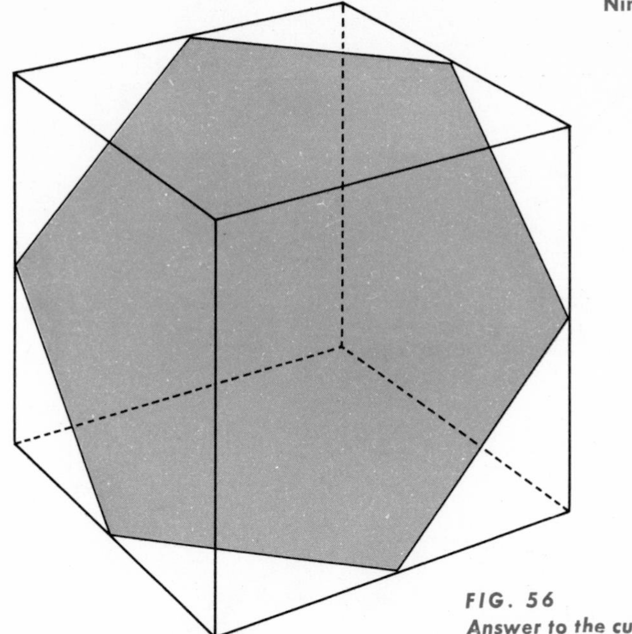

FIG. 56
Answer to the cube-slicing problem.

car, we are told, is $10x$ miles per hour. In $10x$ miles it will pass $60x$ signs, so in one mile it will pass $60x/10x$, or 6, signs. The signs therefore are 1/6 mile, or 880 feet, apart.

4. A cube, cut in half by a plane that passes through the midpoints of six sides as shown in Figure 56, produces a cross section that is a regular hexagon. If the cube is half an inch on the side, the side of the hexagon is $\sqrt{2}/4$ inch.

To cut a torus so that the cross section consists of two intersecting circles, the plane must pass through the center and be tangent to the torus above and below, as shown in Figure 57. If the torus and hole have diameters of three inches and one inch, each circle of the section will clearly have a diameter of two inches.

This way of slicing, and the two ways described earlier, are the only ways to slice a doughnut so that the cross sections are circular. Everett A. Emerson, in the electronics division of National Cash Register, Hawthorne, California, sent a full algebraic proof that there is no fourth way.

5. Figure 58 shows how to construct a straight line that bisects both the Yin and the Yang. A simple proof is obtained by drawing the two broken semicircles. Circle K's diameter is half that of the monad; therefore its area is one-fourth that of the monad.

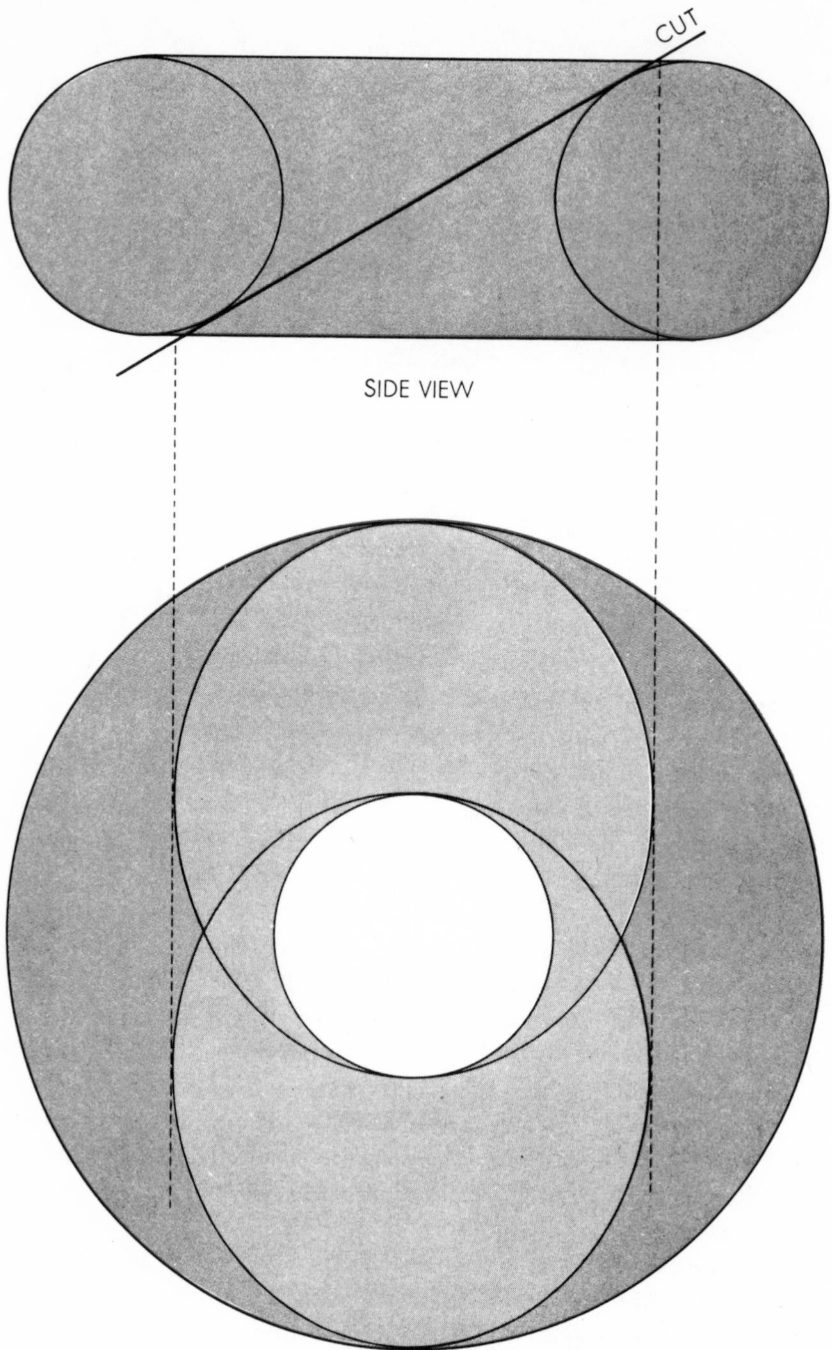

FIG. 57
Answer to the doughnut-slicing problem.

Take region G from this circle, add H, and the resulting region is also one-fourth the monad's area. It follows that area G equals area H, and of course half of G must equal half of H. The bisecting line takes half of G away from circle K, but restores the same area (half of H) to the circle, so the black area below the bisecting line must have the same area as circle K. The small circle's area is one-fourth the large circle's area, therefore the Yin is bisected. The same argument applies to the Yang.

The foregoing proof was given by Henry Dudeney in his answer to problem 158, *Amusements in Mathematics.* After it appeared in *Scientific American,* four readers (A. E. Decae, F. J. Hooven, Charles W. Trigg and B. H. K. Willoughby) sent the following alternative proof, which is much simpler. In Figure 58, draw a horizontal diameter of the small circle K. The semicircle below this line has an area that is clearly 1/8 that of the large circle. Above the diameter is a 45-degree sector of the large circle (bounded by the small circle's horizontal diameter and the diagonal line) which also is obviously 1/8 the area of the large circle. Together, the semicircle and sector have an area of 1/4 that of the large circle, therefore the diagonal line must bisect both Yin and Yang. For ways of bisecting the Yin and Yang with curved lines, the reader is referred to Dudeney's problem, cited above, and Trigg's article, "Bisection of Yin and of Yang," in *Mathematics Magazine,* Vol. 34, No. 2, November-December 1960, pages 107–108.

The Yin-Yang symbol (called the *T'ai-chi-t'u* in China and the *Tomoye* in Japan) is usually drawn with a small spot of Yin inside the Yang and a small spot of Yang inside the Yin. This sym-

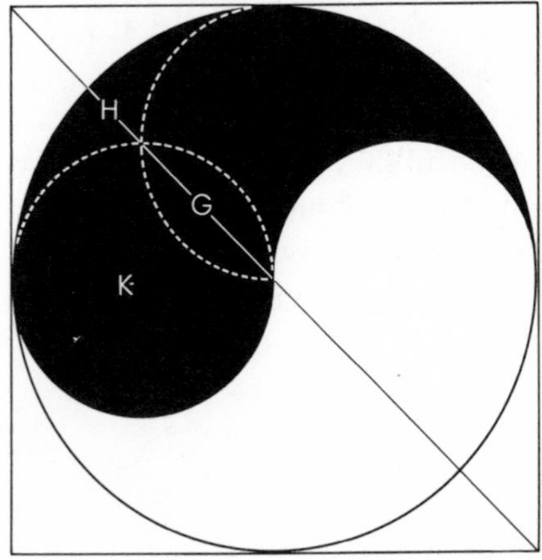

FIG. 58
Answer to the Yin-Yang problem.

bolizes the fact that the great dualities of life are seldom pure; each usually contains a bit of the other. There is an extensive Oriental literature on the symbol. Sam Loyd, who bases several puzzles on the figure (*Sam Loyd's Cyclopedia of Puzzles,* page 26), calls it the Great Monad. The term "monad" is repeated by Dudeney, and also used by Olin D. Wheeler in a booklet entitled *Wonderland,* published in 1901 by the Northern Pacific Railway. Wheeler's first chapter is devoted to a history of the trademark, and is filled with curious information and color reproductions from Oriental sources. For more on the symbol, see Schuyler Cammann, "The Magic Square of Three in Old Chinese Philosophy and Religion," *History of Religions,* Vol. 1, No. 1, Summer 1961, pages 37–80, my *Ambidextrous Universe* (Basic Books, 1965), pages 249–250, and George Sarton, *A History of Science,* Vol. 1 (Harvard University Press, 1952), page 11. Carl Gustav Jung cites some English references on the symbol in his introduction to the book of *I Ching* (1929), and there is a book called *The Chinese Monad: Its History and Meaning,* by Wilhelm von Hohenzollern, the date and publisher of which I do not know.

6. There are probably three blue-eyed Jones sisters and four sisters altogether. If there are n girls, of which b are blue-eyed, the probability that two chosen at random are blue-eyed is:

$$\frac{b(b-1)}{b(n-1)}$$

We are told that this probability is 1/2, so the problem is one of finding integral values for b and n that will give the above expression a value of 1/2. The smallest such values are $n = 4$, $b = 3$. The next highest values are $n = 21$, $b = 15$, but it is extremely unlikely that there would be as many as 21 sisters, so four sisters, three of them blue-eyed, is the best guess.

7. The rose-red city's age is seven billion years. Let x be the city's present age; y, the present age of Time. A billion years ago the city would have been $x - 1$ billion years old and a billion years from now Time's age will be $y + 1$. The data in the problem permit two simple equations:

$$2x = y$$

$$x - 1 = \frac{2}{5}(y + 1)$$

These equations give x, the city's present age, a value of seven billion years; and y, Time's present age, a value of fourteen billion years. The problem presupposes a "Big Bang" theory of the creation of the cosmos.

8. There is space only to suggest the procedure by which it can be shown that Washington High won the high jump event in the track meet involving three schools. Three different positive integers provide points for first, second and third place in each event. The integer for first place must be at least 3. We know there are at least two events in the track meet, and that Lincoln High (which won the shot-put) had a final score of 9, so the integer for first place cannot be more than 8. Can it be 8? No, because then only two events could take place and there is no way that Washington High could build up a total of 22 points. It cannot be 7 because this permits no more than three events, and three are still not sufficient to enable Washington High to reach a score of 22. Slightly more involved arguments eliminate 6, 4 and 3 as the integer for first place. Only 5 remains as a possibility.

If 5 is the value for first place, there must be at least five events in the meet. (Fewer events are not sufficient to give Washington a total of 22, and more than five would raise Lincoln's total to more than 9.) Lincoln scored 5 for the shot-put, so its four other scores must be 1. Washington can now reach 22 in only two ways: 4, 5, 5, 5, 3 or 2, 5, 5, 5, 5. The first is eliminated because it gives Roosevelt a score of 17, and we know that this score is 9. The remaining possibility gives Roosevelt a correct final tally, so we have the unique reconstruction of the scoring shown in the table [*Fig. 59*].

EVENTS	1	2	3	4	5	SCORE
WASHINGTON	2	5	5	5	5	22
LINCOLN	5	1	1	1	1	9
ROOSEVELT	1	2	2	2	2	9

FIG. 59
Answer to the track-meet problem.

Washington High won all events except the shot-put, consequently it must have won the high jump.

Many readers sent shorter solutions than the one just given. Two readers (Mrs. Erlys Jedlicka, Saratoga, California, and Albert Zoch, a student at Illinois Institute of Technology) noticed that there was a short cut to the solution based on the assumption that the problem had a unique answer. Mrs. Jedlicka put it this way:

> *Dear Mr. Gardner:*
>
> *Did you know this problem can be solved without any calculation whatever? The necessary clue is in the last paragraph. The solution to the integer equations must indicate without ambiguity which school won the high jump. This can only be done if one school has won all the events, not counting the shot-put; otherwise the problem could not be solved with the information given, even after calculating the scoring and number of events. Since the school that won the shot-put was not the over-all winner, it is obvious that the over-all winner won the remaining events. Hence without calculation it can be said that Washington High won the high jump.*

9. It is not possible for the termite to pass once through the 26 outside cubes and end its journey in the center one. This is easily demonstrated by imagining that the cubes alternate in color like the cells of a three-dimensional checkerboard, or the sodium and chlorine atoms in the cubical crystal lattice of ordinary salt. The large cube will then consist of 13 cubes of one color and 14 of the other color. The termite's path is always through cubes that alternate in color along the way; therefore if the path is to include all 27 cubes, it must begin and end with a cube belonging to the set of 14. The central cube, however, belongs to the 13 set; hence the desired path is impossible.

The problem can be generalized as follows: A cube of even order (an even number of cells on the side) has the same number of cells of one color as it has cells of the other color. There is no central cube, but complete paths may start on any cell and end on any cell of opposite color. A cube of odd order has one more cell of one color than the other, so a complete path must begin and end on the color that is used for the larger set. In odd-order cubes of orders 3, 7, 11, 15, 19 . . . the central cell belongs to the smaller set, so it cannot be the end of any complete path. In odd-order cubes of 1, 5, 9, 13, 17 . . . the central cell belongs to the larger set, so it can be the end of any path that starts on a cell of the same color. No closed path, going through every unit cube, is possible on any odd-order cube because of the extra cube of one color.

Many two-dimensional puzzles can be solved quickly by similar "parity checks." For example, it is not possible for a rook to start at one corner of a chessboard and follow a path that carries it once through every square and ends on the square at the diagonally opposite corner.

CHAPTER THIRTEEN

□

Polyominoes and Fault-Free Rectangles

POLYOMINOES — the intriguing shapes that cover connected squares on a checkerboard — were introduced to the mathematical world in 1954 by Solomon W. Golomb, now a professor of engineering and mathematics at the University of Southern California. They were first discussed in *Scientific American* in 1957. Since then they have become an enormously popular mathematical recreation, and hundreds of new polyomino puzzles and unusual configurations have come to light. The following communication from Golomb discusses some of these recent discoveries.

"The shapes that cover five connected squares," Golomb writes, "are called pentominoes. There are twelve such shapes. If they are arranged as shown in Figure 60, they resemble letters of the alphabet, and these letters provide convenient names for the pieces. For mnemonic purposes, one has only to remember the end of the alphabet (TUVWXYZ) and the word FILiPiNo.

"In previous articles it was shown that the twelve pentominoes, which have a total of 60 squares, can form such patterns as a three-by-twenty rectangle, a four-by-fifteen rectangle, a five-by-twelve rectangle and a six-by-ten rectangle. They can all be fitted onto the eight-by-eight checkerboard, with the four excess squares of the board forming a two-by-two square at any specified location on the board. Given any pentomino, nine of the others can be used to triplicate it, that is, to form a scale model three

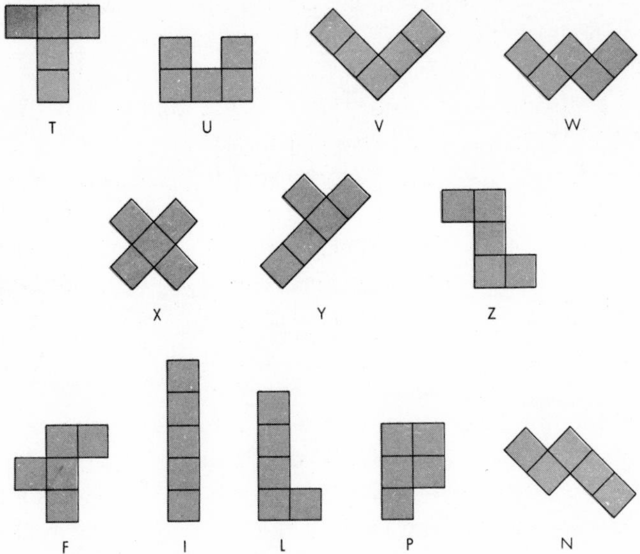

FIG. 60

times as long and three times as high as the selected pentomino. It is also possible to arrange the twelve pentominoes into *two* rectangles, each five by six."

[This last configuration is known as a superposition problem, because it involves shapes that can be superposed. Golomb reports on five new superposition problems, here published for the first time. If the reader has not yet discovered the fascination of playing with pentominoes, he is urged to make a set of them from cardboard and try his skill on some of the puzzles that follow. In all such puzzles, pieces may be placed with either side up.]

"1. Divide the twelve pentominoes into three groups of four each. Find a 20-square shape that each of the three groups will cover. One of several solutions is depicted in Figure 61.

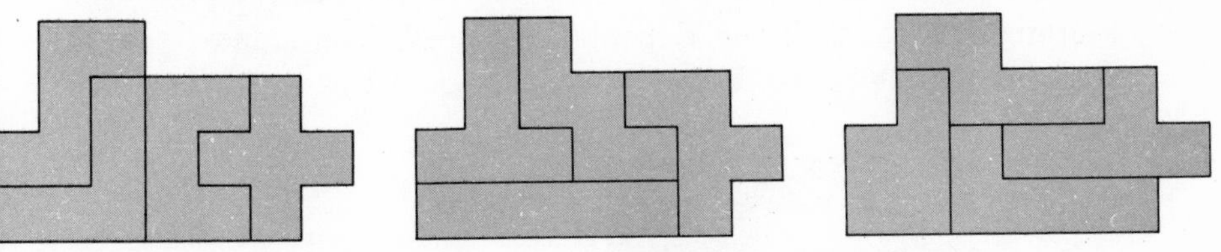

FIG. 61

FIG. 62

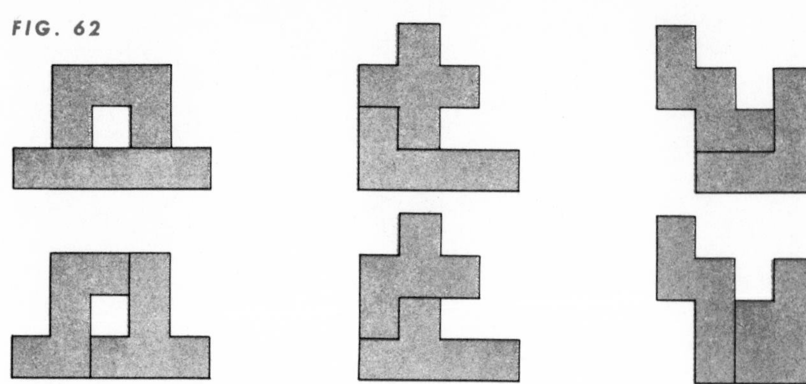

"2. Divide the twelve pentominoes into three groups of four each. Subdivide each group into two pairs of shapes. For each group find a 10-square region that each of the two pairs will cover. One solution is shown in Figure 62. Can the reader find other solutions, including one without holes?

FIG. 63

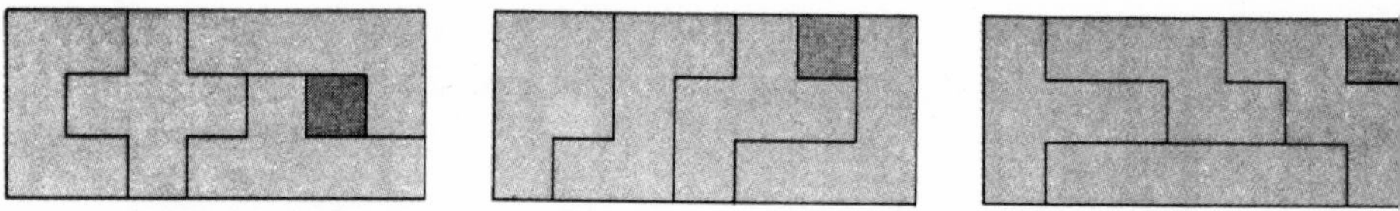

"3. Divide the twelve pentominoes into three groups of four each. To each group add a monomino (a single square), and form a three-by-seven rectangle. Figure 63 shows the solution. It is known to be unique except that in the first rectangle the monomino and Y pentomino can be rearranged and can still occupy the same region.

"The uniqueness proof follows a suggestion by C. S. Lorens. To begin with, in the pattern shown in Figure 64, the X pentomino can be used only in conjunction with the U pentomino. Next, neither the F nor the W pentomino can be used to complete this rectangle. Also, with the U pentomino needed to support the X, it is impossible to use F and W in the *same* three-by-seven rec-

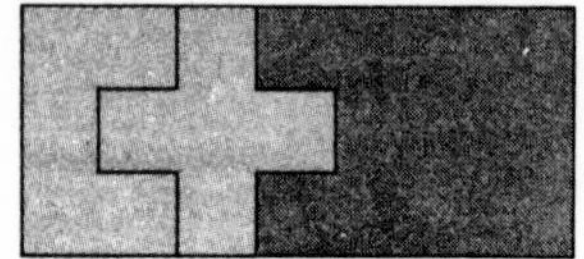

FIG. 64

tangle. Hence, of the three three-by-seven rectangles, one will contain X and U, another will contain W (but not U) and the third will contain F (but not U). When all possible completions of these three rectangles are listed and compared (a very time-consuming enterprise), it is found that the solution shown is the only possible one.

"4. Divide the twelve pentominoes into four groups of three each. Find a 15-square region which each of the four groups will cover. No solution to this problem is known; on the other hand, the problem has not been proved impossible.

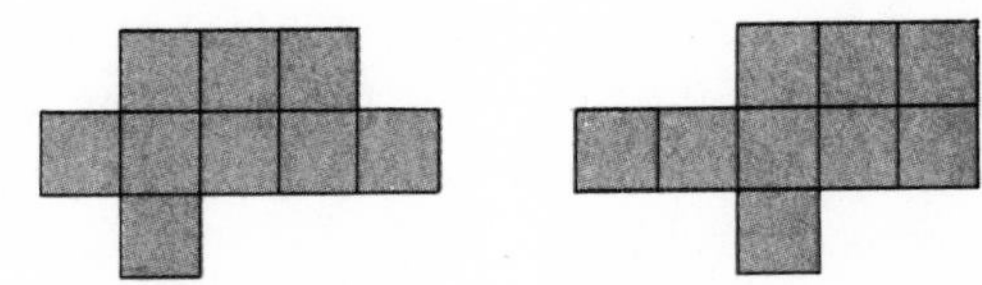

FIG. 65

"5. Find the smallest region on the checkerboard onto which each of the twelve pentominoes, taken one at a time, will fit. The minimum area for such a region is nine squares. There are only two examples of such a region [*Fig. 65*].

"The adequacy of each region is proved by observing that each pentomino in turn will fit on it. The impossibility of fewer than nine squares is proved as follows: If it were possible to use a region with fewer than nine squares, then in particular the I, X and V pentominoes would fit on a region of no more than eight squares. The I and X pentominoes will then have three squares in common. (Otherwise either nine squares are needed, or else the longest straight line has six squares, a needless extravagance.) This can happen in only two distinct ways [*Fig. 66*]. In either case, however, the fitting of the U pentomino would require a ninth square. Thus eight squares are not enough, whereas nine squares have been shown by example to be sufficient.

FIG. 66

"Recently the resources of modern electronic computing have been turned loose on various pentomino problems. The chapter on polyominoes in *The Scientific American Book of Mathematical Puzzles & Diversions* contains a brief account of how Dana S. Scott programed the MANIAC computer at Princeton University for determining all the ways that twelve pentominoes can be fitted onto the eight-by-eight checkerboard, leaving a two-by-two hole in the center. It was discovered that there are 65 basically different solutions in the sense that two solutions differing only by rotation or reflection are not regarded as distinct. More recently, C. B. Haselgrove, an astronomer at the University of Manchester, programed a computer to find all possible ways to form a six-by-ten rectangle with the twelve pentominoes. Excluding rotations and reflections, he found 2,339 basically different solutions! He also verified Scott's program for the eight-by-eight checkerboard problem.

"Several special pentomino configurations make excellent puzzles. Figure 67 shows a 64-square pyramid that can be formed with the twelve pentominoes and the two-by-two square tetromino. The cross in Figure 68 requires only the twelve pentominoes, and is unusually difficult. Still unsolved (neither constructed nor proved impossible) is the pattern shown in Figure 69. Even if the monomino hole is moved to another location, no solution has been found. The closest approximation yet known is pictured in Figure 70. Also believed impossible is Herbert Taylor's configuration, shown in Figure 71, though no one has yet found an impossibility proof.

"Fortunately not all such problems are undecided. The pattern shown in Figure 72, for example, was proved by R. M. Robinson, a mathematician at the University of California, to be incapable of formation by the twelve pentominoes. It has 22 edge squares that form its border. If the pentominoes are examined separately, and the maximum number of edge squares that each could contribute to the pattern are listed, the total proves to be 21, just one short of the required number. This type of reasoning is used in working jigsaw puzzles. It is common practice to separate the edge pieces from the interior pieces so that the picture's border can be made first.

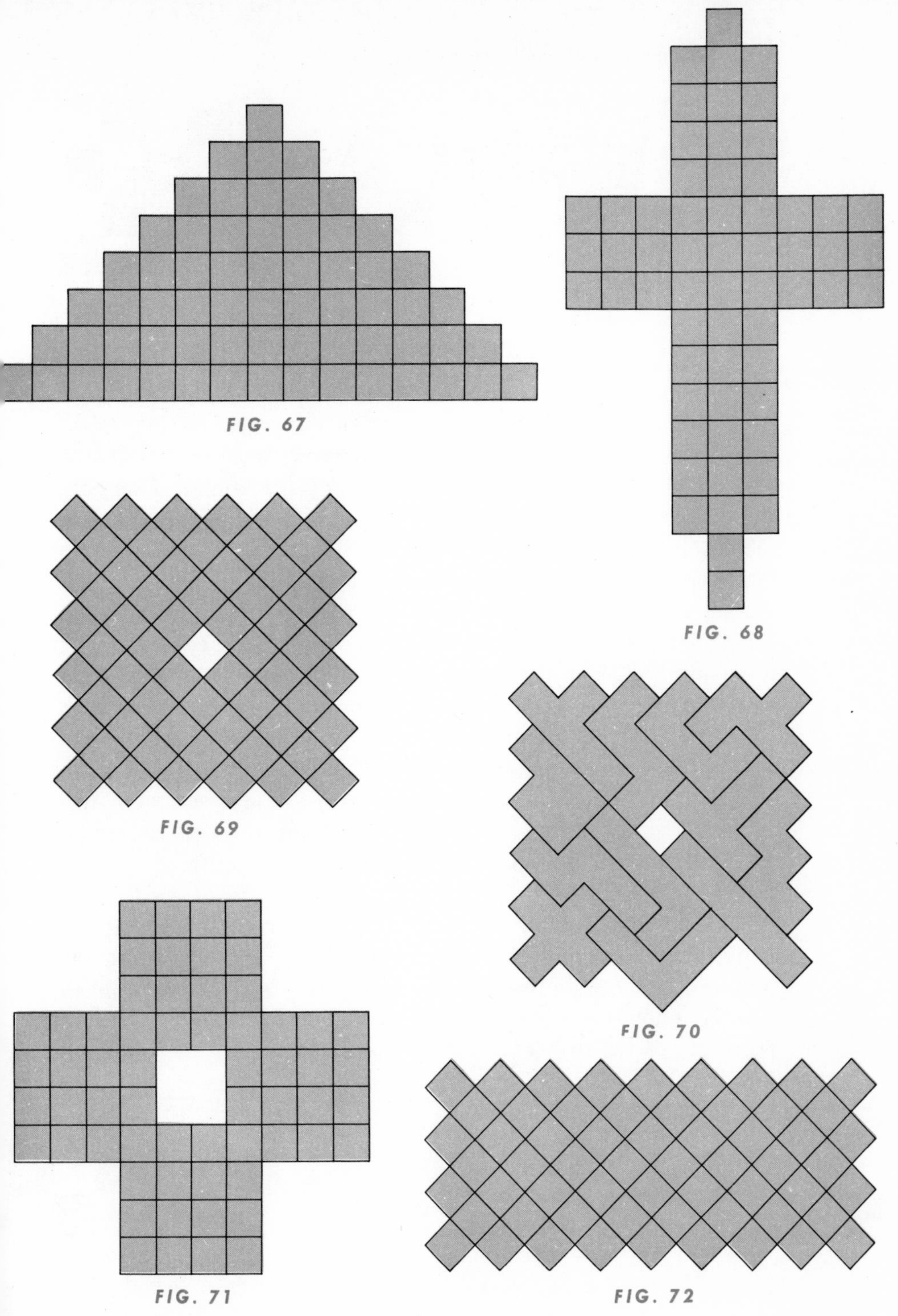

FIG. 67

FIG. 68

FIG. 69

FIG. 70

FIG. 71

FIG. 72

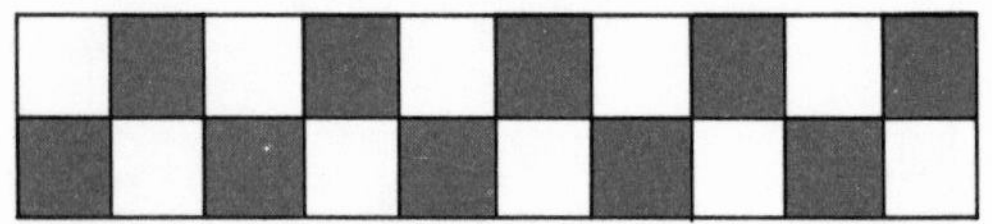
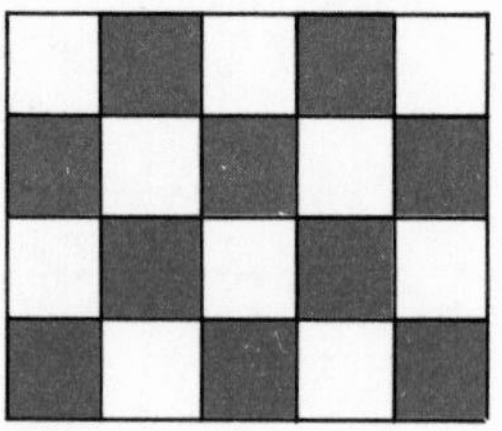

FIG. 73

"Polyominoes that cover four squares of the checkerboard are called tetrominoes. Unlike the pentominoes, the five distinct tetrominoes will not form a rectangle. To prove this, color the squares of a four-by-five rectangle and a two-by-ten rectangle (the only two rectangles with a 20-square area) in checkerboard fashion [*Fig. 73*]. Four of the five tetrominoes [*Fig. 74*] will always cover two dark and two light squares, but the T-shaped tetromino always covers three squares of one color and one square of the other color. Altogether, therefore, the five shapes will cover an odd number of dark squares and an odd number of light squares. However, the two rectangles in question have ten squares of each color, and 10 is an even number.

"On the other hand, any of several different pentominoes can be combined with the five tetrominoes to form a five-by-five square. Two examples are shown in Figure 75. This raises an interesting question: How many different pentominoes can be used in this manner?

"Robert I. Jewett, a graduate student in mathematics at the University of Oregon (he was mentioned in the answer to the first problem of the previous chapter), has proposed a problem involving dominoes (2-square polyominoes) that is quite different from any of the problems just discussed. Is it possible to form a

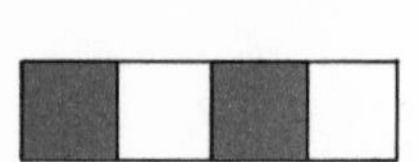
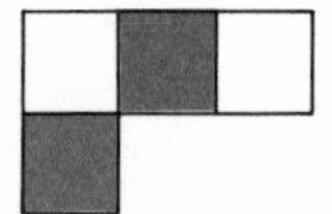
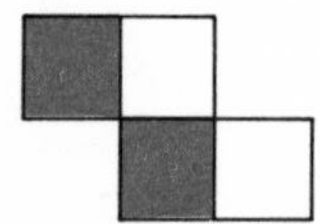
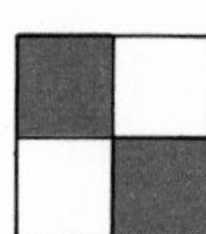
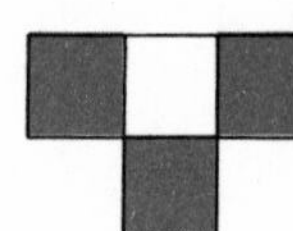

FIG. 74

rectangle with dominoes in such a way that there is no straight line, vertical or horizontal, that joins opposite sides of the rectangle? For example, in Figure 76 there is a vertical line in the center that extends all the way from top to bottom. If dominoes are thought of as bricks, such a line represents a structural weakness. Jewett's problem is thus one of finding rectangular masonry patterns without 'fault lines.' Many people who try this problem soon give up, convinced that there are no solutions. Actually, there are infinitely many."

The reader is invited to make or obtain a set of dominoes — the standard set of 28 dominoes is more than sufficient — and see if he can determine the smallest possible "fault free" rectangle that can be made with them. The solution to this beautiful problem will be given in the answer section, together with a remarkable proof, devised by Golomb, that there are no fault-free six-by-six squares.

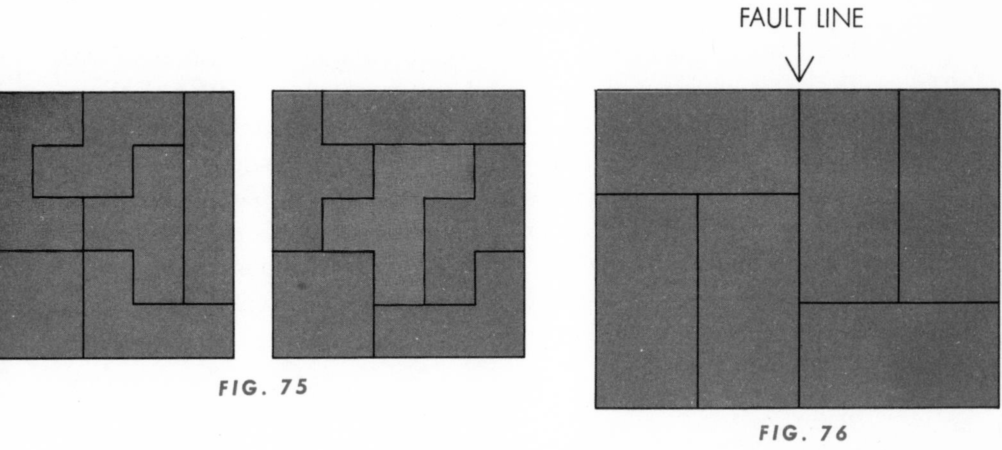

FIG. 75

FIG. 76

ADDENDUM

SINCE THIS chapter appeared in *Scientific American,* much progress has been made in the study of polyominoes and fault-free rectangles. The interested reader is urged to look into Golomb's book *Polyominoes,* published by Scribner's in 1965, in which the field is thoroughly covered and many new results given.

The Herbert Taylor configuration [*Fig. 71*] and the jagged square [*Fig. 69*] have both been proved impossible, though no short, elegant proofs have yet been found for either pattern. On the Taylor configuration I received proofs from Ivan M. Anderson, Leo J. Brandenburger, Bruce H. Douglas, Micky Earnshaw, John G. Fletcher, Meredith G. Williams and Donald L. Vanderpool. Impossibility proofs for the jagged square came from Bruno Antonelli, Leo J. Brandenburger, Cyril B. Carstairs, Bruce H. Douglas, Micky Earnshaw, E. J. Mayland, Jr., and Robert Nelson.

J. A. Lindon, Surrey, England, found a solution of the jagged square with the monomino (hole) on the border, adjacent to a corner (his solution appears on page 73 of Golomb's book). Other readers found solutions with the monomino at the corner. D. C. and B. G. Gunn, of Sussex, England, sent sixteen different patterns of this type. It is not yet known if the monomino can be on the border and next to the corner but one.

William E. Patton, a retired hydraulics engineer living in South Boston, Virginia, wrote that he had been investigating fault-free domino rectangles since 1944. He sent me summaries of some of his results, many of them suggesting interesting problems. What, for instance, is the smallest fault-free rectangle with the same number of horizontal and vertical dominoes? The answer is the five-by-eight. Readers may like to search for solutions.

The concept of the fault-free domino square suggests a variety of games, none of which, as far as I know, have been investigated. For example, players take turns placing dominoes on a square, checked board. The winner is the first to complete a fault line, either vertically or horizontally. Or the game can be played in reverse: The first to complete a fault line loses.

ANSWERS

ANSWERS to the pyramid and cross puzzles are depicted in Figures 77 and 78. Neither solution is unique. Readers were asked to determine which individual pentominoes can be combined with the five tetrominoes to form a five-by-five square. This is possible with all pentominoes except the I, T, X and V.

The smallest fault-free rectangle (a rectangle with no straight line joining opposite sides) that can be formed with dominoes is a five-by-six. The two basically different solutions are shown in Figure 79.

FIG. 77
An answer to the pyramid puzzle.

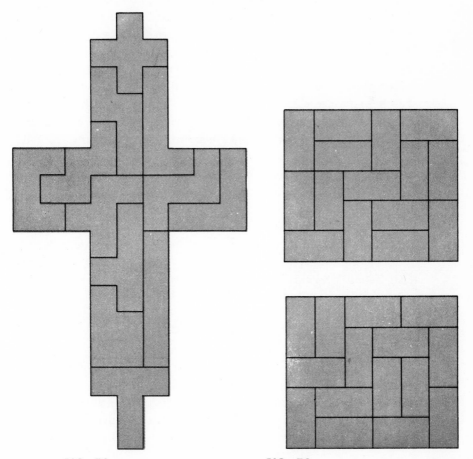

FIG. 78
An answer to the cross puzzle.

FIG. 79
Answers to the fault-free rectangle puzzle.

"It is not difficult to show," writes Solomon W. Golomb, "that the minimum width for fault-free rectangles must exceed 4. (Cases of width 2, 3 and 4 are best treated separately.) Therefore, since five-by-five is an odd number of squares, and dominoes always cover an even number of squares, the five-by-six rectangle is the smallest solution.

"A five-by-six rectangle can be extended to an eight-by-eight checkerboard and still satisfy the fault-free condition. An example is shown in Figure 80. Surprisingly, there are no fault-free six-by-six rectangles. For this there is a truly remarkable proof.

"Imagine any six-by-six rectangle covered entirely with dominoes. Such a figure contains eighteen dominoes (half the area) and ten grid lines (five horizontal and five vertical). It is fault-free if each grid line intersects at least one domino.

"The first step in our proof is to show that in any fault-free rectangle each grid line must cut an *even* number of dominoes. Consider any vertical grid line. The area to the left of it (expressed in number of unit squares) is even (6, 12, 18, 24 or 30). Dominoes *entirely* to the left of this grid line must cover an even area because each domino covers two squares. Dominoes *cut* by the grid line must also occupy an even area to the left of it, because this area is the difference between two even numbers (the total area to the left, and the area of the uncut dominoes to the left). Since each cut domino occupies *one* square to the left of the grid line, there must be an *even* number of dominoes cut by the grid line.

"The six-by-six square has ten grid lines. To be fault-free, each line must intersect at least two dominoes. No domino can be cut by more than one grid line, therefore at least twenty dominoes must be cut by grid lines. But there are only eighteen dominoes in a six-by-six square!

"Similar reasoning shows that for a fault-free six-by-eight rectangle to exist, every grid line must intersect *exactly* two dominoes. Such a rectangle is shown in Figure 81.

"The most general result is the following: If a rectangle has even area, and both its length and width exceed 4, it is possible to find a fault-free covering of the rectangle with dominoes, *except* in the case of the six-by-six square. Actually, coverings for all larger rectangles can be obtained from the five-by-six rectangle and the six-by-eight, using a method of enlarging either the length or width by 2. This method is easiest to explain by Figure 82. To extend it horizontally by 2, a horizontal domino is placed next to each horizontal domino at the old boundary, while vertical dominoes are shifted from the old boundary to the new, with the intervening space filled by two horizontal dominoes.

"The reader may find it interesting to study trominoes as bricks. In particular, what is the smallest rectangle that can be covered by two or more 'straight trominoes' (one-by-three rectangles) without any fault lines?"

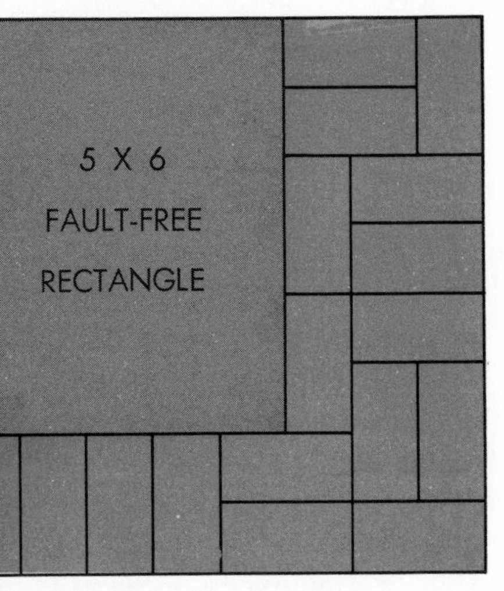

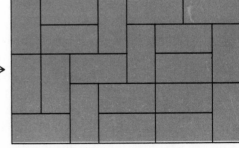

. 80
ault-free rectangle on an eight-by-eight board.

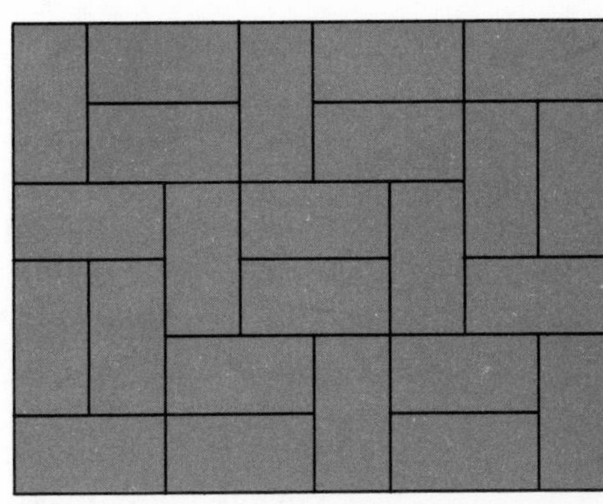

FIG. 81
A fault-free six-by-eight rectangle.

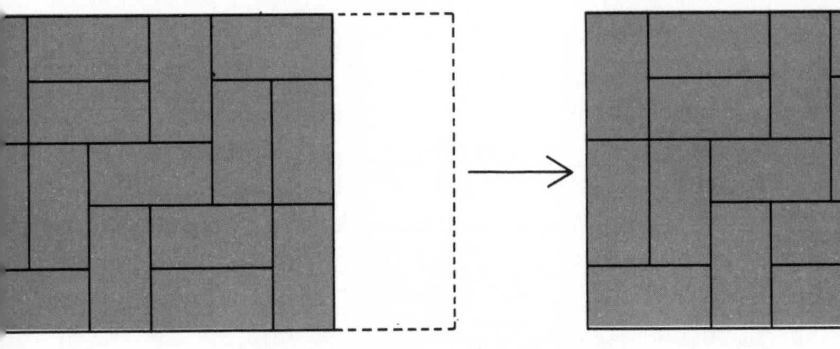

FIG. 82
A general solution to the fault-free rectangle puzzle.

CHAPTER FOURTEEN

□

Euler's Spoilers: The Discovery of an Order-10 Graeco-Latin Square

THE HISTORY of mathematics is filled with shrewd conjectures — intuitive guesses by men of great mathematical insight — that often wait for centuries before they are proved or disproved. When this finally happens, it is a mathematical event of first magnitude. Not one but two such events were announced in April 1959 at the annual meeting of the American Mathematical Society. We need not be concerned with one of them (a proof of a conjecture in advanced group-theory), but the other, a disproof of a famous guess by the great Swiss mathematician Leonhard Euler (pronounced "oiler"), is related to many classical problems in recreational mathematics. Euler had expressed his conviction that Graeco-Latin squares of certain orders could not exist. Three mathematicians (E. T. Parker, of Remington Rand Univac, a division of the Sperry Rand Corporation, and R. C. Bose and S. S. Shrikhande, of the University of North Carolina) completely demolished Euler's conjecture. They have found methods for constructing an infinite number of squares of the type that experts, following Euler, for 177 years had believed to be impossible.

The three mathematicians, dubbed "Euler's spoilers" by their colleagues, have written a brief account of their discovery. The following quotations from this account are interspersed with comments of my own to clarify some of the concepts or to summarize the more technical passages.

"In the last years of his life Leonhard Euler (1707–1783) wrote a lengthy memoir on a new species of magic square: *Recherches sur une nouvelle espèce de quarres magiques*. Today these constructions are called Latin squares after Euler's practice of labeling their cells with ordinary Latin letters (as distinct from Greek letters).

"Consider, for example, the square at the left in Figure 83. The four Latin letters a, b, c and d occupy the sixteen cells of the square in such a way that each letter occurs once in every row and once in every column. A different Latin square, its cells labeled with the four corresponding Greek letters, is shown in the middle of the illustration. If we superpose these two squares, as shown at the right, we find that each Latin letter combines once and only once with each Greek letter. When two or more Latin squares can be combined in this way, they are said to be orthogonal squares. The combined square is known as a Graeco-Latin square."

The square at the right provides one solution to a popular card puzzle of the 18th century: Take all the aces, kings, queens and jacks from a deck and arrange them in a square so that every row and column will contain all four values and all four suits. Readers may enjoy searching for another solution in which the two main diagonals also show one of each suit and one of each value.

"In general a Latin square of order n is defined as an n-by-n square, the n^2 cells of which are occupied by n distinct symbols, such that each symbol occurs exactly once in each row and once

a	b	c	d
b	a	d	c
c	d	a	b
d	c	b	a

α	β	γ	δ
γ	δ	α	β
δ	γ	β	α
β	α	δ	γ

aα	bβ	cγ	dδ
bγ	aδ	dα	cβ
cδ	dγ	aβ	bα
dβ	cα	bδ	aγ

FIG. 83
***The Graeco-Latin square* (right) *is formed by superposing two Latin squares* (left and center).**

0	1	2	3	4
1	2	3	4	0
2	3	4	0	1
3	4	0	1	2
4	0	1	2	3

0	1	2	3	4
2	3	4	0	1
4	0	1	2	3
1	2	3	4	0
3	4	0	1	2

0	1	2	3	4
3	4	0	1	2
1	2	3	4	0
4	0	1	2	3
2	3	4	0	1

0	1	2	3	4
4	0	1	2	3
3	4	0	1	2
2	3	4	0	1
1	2	3	4	0

FIG. 84
Four mutually orthogonal Latin squares of order 5.

in each column. There may exist a set of two or more Latin squares such that any pair of them is orthogonal. In Figure 84 are shown four mutually orthogonal Latin squares of order 5, which use digits for their symbols."

In Euler's day it was easy to prove that no Graeco-Latin square of order 2 is possible. Squares of orders 3, 4 and 5 were known, but what about order 6? Euler put it this way: Each of six different regiments has six officers, one belonging to each of six different ranks. Can these 36 officers be arranged in a square formation so that each row and file contains one officer of each rank and one of each regiment?

"Euler showed that the problem of n^2 officers, which is the same as the problem of constructing a Graeco-Latin square of order n, can always be solved if n is odd, or if n is an 'evenly even' number (that is, a number divisible by 4). On the basis of extensive trials he stated: 'I do not hesitate to conclude that it is impossible to produce any complete square of 36 cells, and the same possibility extends to the cases of $n = 10$, $n = 14$ and in general to all unevenly even numbers' (even numbers not divisible by 4). This became famous as Euler's conjecture. It may be stated more formally as follows: There does not exist a pair of orthogonal Latin squares of order $n = 4k + 2$ for any positive integer k."

In 1901 the French mathematician Gaston Tarry published a proof that Euler's conjecture did indeed hold for a square of order 6. Tarry, assisted by his brother, did it the hard way. He simply listed all the possible ways of constructing an order-6 Latin square, then showed that no pair would form a Graeco-Latin square. This, of course, strengthened Euler's conjecture. Several mathematicians even published "proofs" that the conjecture was true, but the proofs were later found to contain flaws.

The labor involved in settling the question by exhaustive pencil-and-paper enumeration goes up rapidly as the order of the square increases. The next unknown case, the order 10, was far too complex to be settled in this way, and in 1959 it was almost beyond the range of computers. At the University of California at Los Angeles, mathematicians programed the SWAC computer to search for order-10 Graeco-Latin squares. After more than 100 hours of running time, it failed to find a single one. The search was confined, however, to such a microscopic portion of the total possible cases that no conclusion could be drawn. It was estimated that, if Euler's conjecture were true, it would take the fastest 1959 computer, using the program SWAC had used, at least a century to prove it.

"The last sentence of Euler's memoir reads: 'At this point I close my investigations on a question, which though of little use in itself, led us to rather important observations for the doctrine of combinations, as well as for the general theory of magic squares.' It is a striking example of the unity of science that the initial impulse which led to a solution of Euler's conjecture came from the practical needs of agricultural experimentation, and that the investigations which Euler thought useless have proved to have enormous value in the design of controlled experiments."

Sir Ronald Fisher, now professor of genetics at the University of Cambridge and one of the world's leading statisticians, was the first to show (in the early 1920's) how Latin squares could be used in agricultural research. Suppose, for example, one wishes to test with a minimum waste of time and money the effects of seven different agricultural chemicals on the growth of wheat. One difficulty encountered in such a test is that the fertility of different patches of soil usually varies in an irregular way. How can we design an experiment that will simultaneously test all seven chemicals and at the same time eliminate any "bias" due to these fertility variations? The answer: Divide the wheat field into "plots" that are the cells of a 7-by-7 square, then apply the seven "treatments" in the pattern of a randomly chosen Latin square. Because of the pattern a simple statistical analysis of the results will eliminate any bias due to variations in soil fertility.

Suppose that instead of one variety of wheat for this test we have seven. Can we design an experiment that will take this fourth variable into account? (The other three variables are row fertility, column fertility and type of treatment.) The answer is now a Graeco-Latin square. The Greek letters show where to plant the seven varieties of wheat and the Latin letters where to apply the seven different chemicals. Again the statistical analysis of results is simple.

Graeco-Latin squares are now widely used for designing experiments in biology, medicine, sociology and even marketing. The "plot" need not, of course, be a piece of land. It may be a cow, a patient, a leaf, a cage of animals, the spot where an injection is made, a period of time or even an observer or group of observers. The Graeco-Latin square is simply the chart of the experiment. Its rows take care of one variable, columns take care of another, the Latin symbols a third and the Greek symbols a fourth. For example, a medical investigator may wish to test the effects of five different types of pill (one a placebo) on persons in five different age brackets, five different weight groups and five different stages of the same disease. A Graeco-Latin square of order 5, selected randomly from all possible squares of that order, is the most efficient design the investigator can use. More variables can be accommodated by superposing additional Latin squares, though for any order n there are never more than $n - 1$ squares that are mutually orthogonal.

The story of how Parker, Bose and Shrikhande managed to find Graeco-Latin squares of orders 10, 14, 18, 22 (and so on) begins in 1958, when Parker made a discovery that cast grave doubt on the correctness of Euler's conjecture. Following Parker's lead, Bose developed some strong general rules for the construction of large-order Graeco-Latin squares. Then Bose and Shrikhande, applying these rules, were able to construct a Graeco-Latin square of order 22. Since 22 is an even number not divisible by 4, Euler's conjecture was contradicted. It is interesting to note that the method of constructing this square was based on the solution of a famous problem in recreational mathematics called Kirkman's schoolgirl problem, proposed by T. P. Kirkman in 1850. A schoolteacher is in the habit of taking her fifteen girls for a daily walk, always arranging them three abreast in five rows. The problem is to arrange them so that for seven consecutive days no girl will walk more than once in the same row with any other girl. The solution to this problem is an example of an important type of experimental design known as "balanced incomplete blocks."

When Parker saw the results obtained by Bose and Shrikhande, he was able to develop a new method that led to his construction of an order-10 Graeco-Latin square. It is shown in Figure 85. The symbols of one Latin square are the digits 0 to 9 on the left side of each cell. The digits on the right side of each cell belong to the second Latin square. With the aid of this square, the very existence of which is denied in many current college textbooks on experimental methods, statisticians can now design for the first time experiments in which four sets of variables, each with ten different values, can be kept easily and efficiently under control.

(Note that the order-3 square at the lower right corner of the order-10 square is an order-3 Graeco-Latin square. All order-10 squares that were first constructed by Parker and his collaborators contained an order-3 subsquare in the sense that one could always form the smaller square by permuting the rows and columns of the larger one. Changing the order of rows or columns obviously does not affect the properties of any Graeco-Latin square. Such permutations are trivial; if one square can be obtained from another by shifting rows or columns, the two squares are considered the "same" square. For a while it was an open question whether all order-10 Graeco-Latin squares possessed order-3 subsquares, but this conjecture was shown false when many squares were discovered that did not have this feature.)

"At this stage," the three mathematicians conclude their report, "there ensued a feverish correspondence between Bose and Shrikhande on the one hand and Parker on the other. Methods were refined more and more; it was ultimately established that Euler's conjecture is wrong for *all* values of $n = 4k + 2$, where n is greater than 6. The suddenness with which complete success came in a problem that had baffled mathematicians for almost two centuries startled the authors as much as anyone else. What makes this even more surprising is that the concepts employed were not even close to the frontiers of deep modern mathematics."

00	47	18	76	29	93	85	34	61	52
86	11	57	28	70	39	94	45	02	63
95	80	22	67	38	71	49	56	13	04
59	96	81	33	07	48	72	60	24	15
73	69	90	82	44	17	58	01	35	26
68	74	09	91	83	55	27	12	46	30
37	08	75	19	92	84	66	23	50	41
14	25	36	40	51	62	03	77	88	99
21	32	43	54	65	06	10	89	97	78
42	53	64	05	16	20	31	98	79	87

FIG. 85
E. T. Parker's Graeco-Latin square of order 10, a counter-example to Euler's conjecture.

ADDENDUM

In the years following 1959, computer speeds increased enormously, as well as the ingenuity of mathematicians in devising more efficient methods of programing. Using a technique called "backtrack," Parker planned a program for the UNIVAC 1206 Military Computer that was able to take a given order-10 Latin square and complete an exhaustive search for an orthogonal companion in from 28 to 45 minutes of running time. This improved on the search time of the old SWAC program by a factor of about one trillion! Result: the production of hundreds of new Graeco-Latin squares of order 10. Indeed, it turned out that such squares are quite common. UNIVAC found orthogonal mates for more than half of the randomly constructed order-10 Latin squares that were fed to it. "Thus Euler guessed wrong by a large margin," Parker has written, "and the evidence from early computation demonstrated only that the search is of large magnitude."

The big disappointment in the recent computer work on Graeco-Latin squares is that, so far, no triplet of mutually orthogonal Latin squares of order 10 has been found. It had earlier been proved that for any order n, the largest possible number of mutually orthogonal Latin squares is $n - 1$. A set of $n - 1$ such squares is known as a "complete set." For example, the order-2 Latin square has a complete set that consists of the single square itself. The order-3 square has a complete set of two orthogonal squares, and the order-4 square has a complete set of three. A complete set of four mutually orthogonal Latin squares of order 5 is shown in Fig. 84. (Any pair of these will, of course, combine to produce a Graeco-Latin square.) No complete set of order 6 exists; indeed, not even a pair. Complete sets do exist for orders 7, 8 and 9. Order 10, therefore, is the lowest for which it is not yet known if a complete set is possible. It is not even known if a set of three exists.

The question takes on added interest because of its connection with what are called "finite projective planes." (The interested reader will find these fascinating structures explained in several of the references listed in the bibliography for this chapter.) It has been shown that if a complete set of mutually orthogonal Latin squares exists for a given order n, it is possible to derive from it a construction of a finite projective plane of order n. Conversely, if a finite projective plane is known for order n, one can construct a complete set of mutually orthogonal Latin squares of order n. Since Tarry showed that not even two orthogonal Latin squares of order 6 are possible, it follows that no finite projective plane of order 6 is possible. Complete sets (and hence finite projective planes) exist for orders 2, 3, 4, 5, 7, 8 and 9. The lowest-order finite projective plane, the existence of which has been neither proved nor disproved, is order 10. The discovery, therefore, of a complete set of nine order-10 Latin squares would simultaneously answer a major unsolved problem about finite projective planes. At the moment, the question is beyond the scope of computer programs, and not likely to be solved unless computer speeds greatly increase or someone discovers a new approach that leads to a breakthrough.

Scientific American's cover for November 1959 reproduced a striking oil painting by the magazine's staff artist, Emi Kasai, showing the order-10 Graeco-Latin square that is given here in Figure 85. The ten digits were replaced by ten different colors, so that each cell contained a unique pair of colors. Fig. 86 shows a handsome needlepoint rug made in 1960, by Mrs. Karl Wihtol,

FIG. 86
A needlepoint rug based on Parker's Graeco-Latin square.

FIG. 87
A solution to card problem.

Middletown, New Jersey, that duplicates the cover painting. (It is equivalent to the square in Figure 85 after it has been given a clockwise quarter-turn.) The outside colors of each cell form one Latin square, the inside colors form the other. In every row and column each color appears only once as an outside color and only once as an inside color. Miss Kasai's original painting was purchased by Remington Rand and presented as a gift to Parker.

ANSWERS

FIGURE 87 shows one way of arranging the sixteen highest playing cards so that no value or suit appears twice in any row, column or the two main diagonals. Note that the four cards at each corner, as well as the four central cards, also form sets in which each value and suit are represented. It would be nice if a solution also permitted the colors to be alternated checkerboard fashion, but this is not possible.

W. W. Rouse Ball, in *Mathematical Recreations and Essays* (current edition, page 190), cites a 1723 source for the problem and says that it has 72 fundamentally different solutions, not counting rotations and reflections as different. But Henry Ernest Dudeney, in *Amusements in Mathematics,* problem 304, traces the puzzle back to a 1624 edition of a book by Claude Gaspar Bachet, and points out the error in the computation of 72 different solutions. There are 144. This was independently worked out by Bernard Goldenberg, of Brooklyn, after I had given the incorrect figure in my answer.

If only rows and columns are considered (and the two main diagonals ignored), it *is* possible to find solutions in which the colors alternate like a checkerboard. Adolf Karfunkel, of New York City, sent me several such solutions of which the following is one:

QH	KC	JD	AS
JC	AH	QS	KD
AD	JS	KH	QC
KS	QD	AC	JH

CHAPTER FIFTEEN

□

The Ellipse

A circle no doubt has a certain appealing simplicity at the first glance, but one look at an ellipse should have convinced even the most mystical of astronomers that the perfect simplicity of the circle is akin to the vacant smile of complete idiocy. Compared to what an ellipse can tell us, a circle has little to say. Possibly our own search for cosmic simplicities in the physical universe is of this same circular kind — a projection of our uncomplicated mentality on an infinitely intricate external world.

— Eric Temple Bell,
Mathematics: Queen and Servant of Science

MATHEMATICIANS have a habit of studying, just for the fun of it, things that seem utterly useless; then centuries later their studies turn out to have enormous scientific value. There is no better example of this than the work done by the ancient Greeks on the noncircular curves of second degree: the ellipse, parabola and hyperbola. They were first studied by one of Plato's pupils. No important scientific applications were found for them until the 17th century, when Kepler discovered that planets move in ellipses and Galileo proved that projectiles travel in parabolas.

Apollonius of Perga, a third century B.C. Greek geometer, wrote the greatest ancient treatise on these curves. His work *Conics* was the first to show how all three curves, along with the circle, could be obtained by slicing the same cone at continuously varying angles. If a plane is passed through a cone so that it is parallel to the base [*see Fig. 88*], the section is a circle. If the plane is tipped, no matter how slightly, the section becomes elliptical. The more the plane is tipped, the more elongated the ellipse becomes, or, as

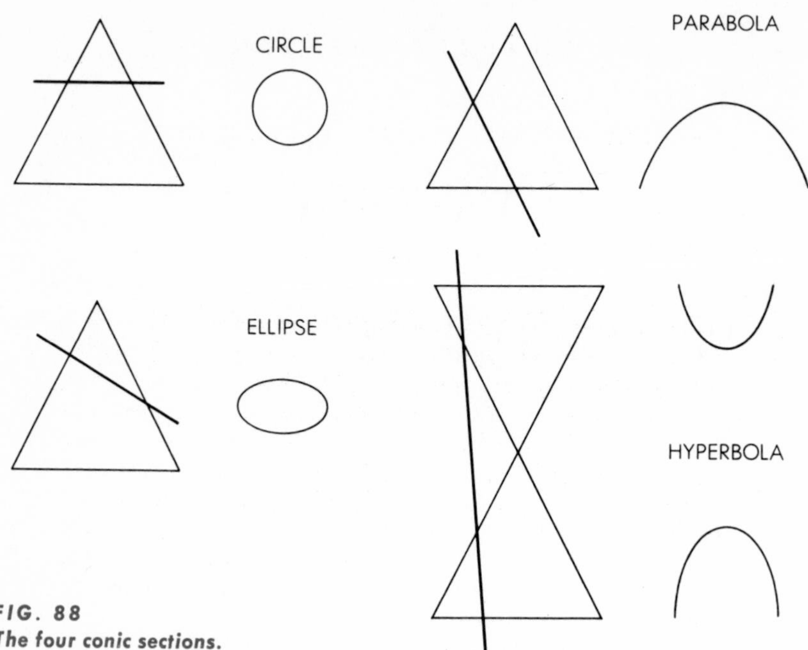

FIG. 88
The four conic sections.

the mathematician puts it, the more eccentric. One might expect that as the plane became steeper the curve would take on more of a pear shape (since the deeper the slice goes, the wider the cone), but this is not the case. It remains a perfect ellipse until the plane becomes parallel to the side of the cone. At this instant the curve ceases to close on itself; its arms stretch out toward infinity and the curve becomes a parabola. Further tipping of the plane causes it to intersect an inverted cone placed above the other one [*see bottom of Fig. 88*]. The two conic sections are now the two branches of a hyperbola. (It is a common mistake to suppose that the plane must be parallel to the cone's axis to cut a hyperbola.) They vary in shape as the cutting plane continues to rotate until finally they degenerate into straight lines. The four curves are called curves of second degree because they are the Cartesian graph forms of all second-degree equations that relate two variables.

The ellipse is the simplest of all plane curves that are not straight lines or circles. It can be defined in numerous ways, but perhaps the easiest to grasp intuitively is this: An ellipse is the locus, or path, of a point moving on a plane so that the sum of its distances from two fixed points is constant. This property underlies a well-known method of drawing an ellipse. Stick two thumbtacks in a sheet of paper, put a loop of string around them and keep the string taut with the point of a pencil as shown in Figure 89. Moving the pencil around the tacks will trace a perfect ellipse. (The length of the cord cannot vary; therefore the sum of the distances of the pencil point from the two tacks remains constant.) The two fixed points (tacks) are called the foci of the ellipse. They lie on the major axis. The diameter perpendicular to this axis is the minor axis. If you move the tacks closer together (keeping the same loop of cord), the ellipse becomes less and less eccentric. When the two foci come together, the ellipse becomes a circle. As the foci move farther apart, the ellipse becomes more elongated until it finally degenerates into a straight line.

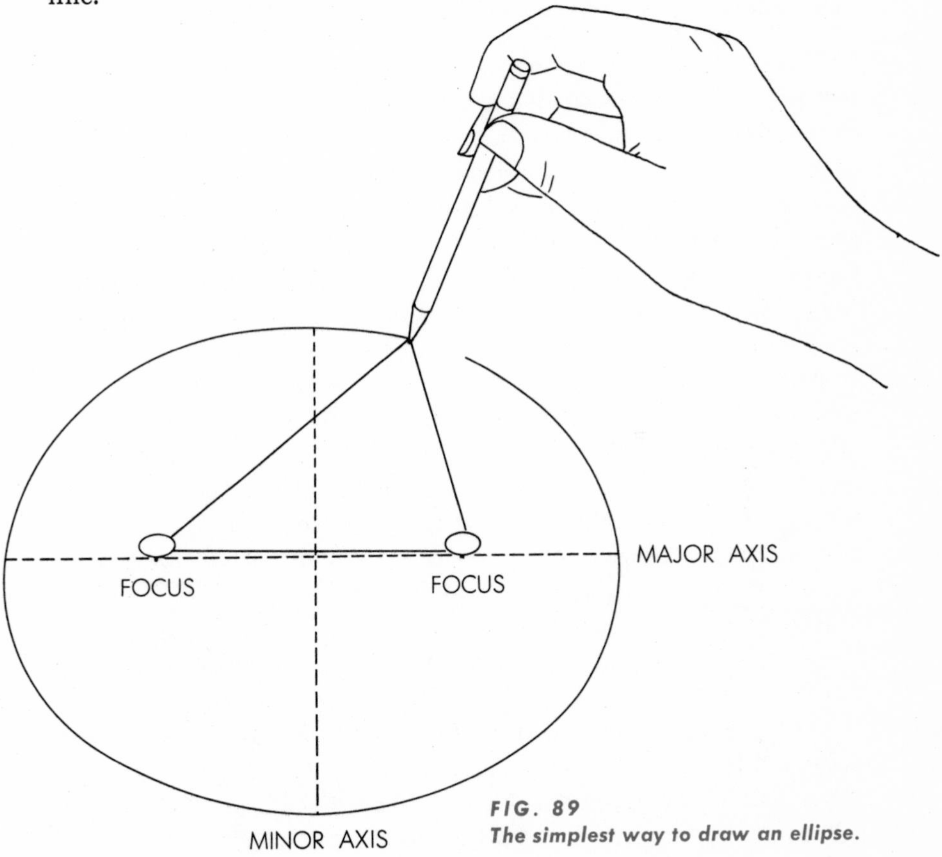

FIG. 89
The simplest way to draw an ellipse.

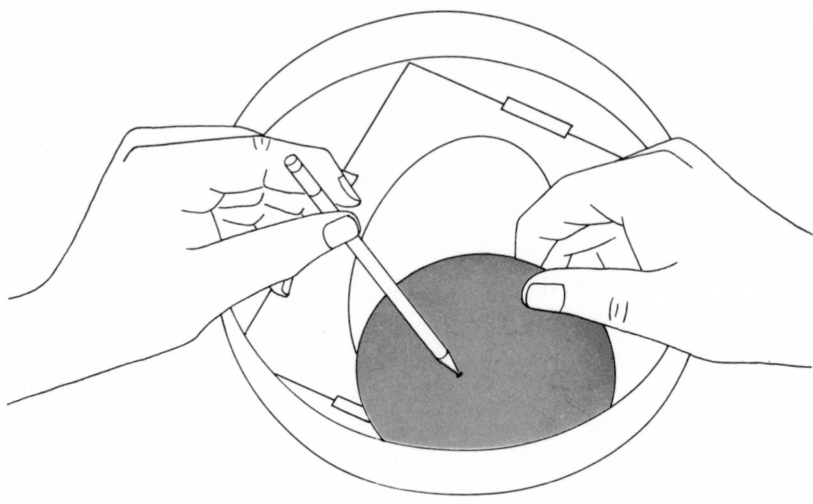

FIG. 90
An ellipsograph made with a circular cake pan and a cardboard disk.

There are many other ways to construct ellipses. One curious method can be demonstrated with a circular cake pan and a cardboard disk having half the diameter of the pan. Put friction tape or masking tape around the inside rim of the pan to keep the disk from slipping when it is rolled around the rim. Anchor a sheet of paper to the bottom of the pan with strips of cellophane tape at the edges. Punch a hole anywhere in the disk with a pencil, place the point of the pencil on the paper and roll the disk around the pan [*see Fig. 90*]. An ellipse will be drawn on the paper. It is best to hold the pencil lightly with one hand while turning the disk slowly with the other, keeping it pressed firmly against the rim of the pan. If the hole is at the center of the disk, the pencil point will of course trace a circle. The nearer the hole is to the edge of the disk, the greater the eccentricity of the ellipse will be. A point on the circumference of the disk traces an ellipse that has degenerated into a straight line.

Here is another pleasant way to obtain an ellipse. Cut a large circle from a sheet of paper. Make a spot somewhere inside the circle, but not at the center, then fold the circle so that its circumference falls on the spot. Unfold, then fold again, using a different point on the circumference, and keep repeating this until the paper has been creased many times in all directions. The creases form a set of tangents that outline an ellipse [*see Fig. 91*].

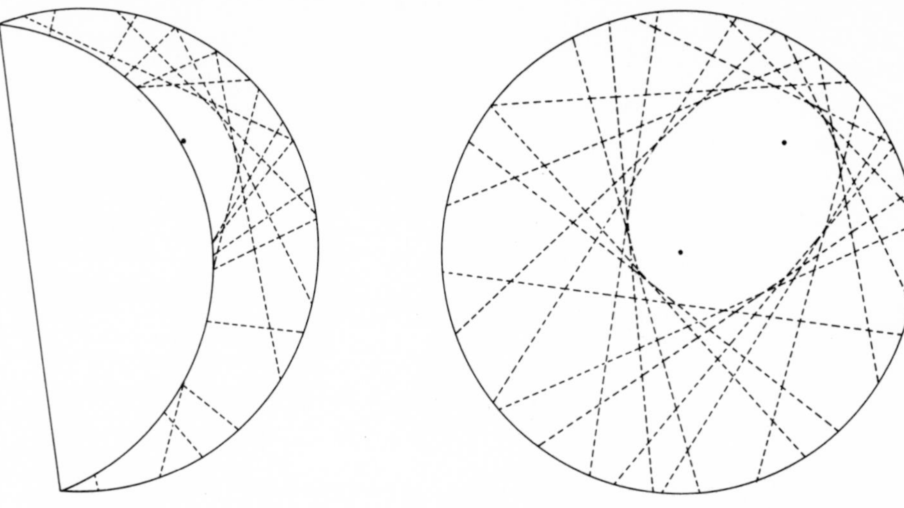

FIG. 91
Folding a paper circle so that its edge falls on an off-center spot makes an ellipse.

Though not so simple as the circle, the ellipse is nevertheless the curve most often "seen" in everyday life. The reason is that every circle, viewed obliquely, appears elliptical. In addition, all closed noncircular shadows cast on a plane by circles and spheres are ellipses. Shadows on the sphere itself — the inner curve of a crescent moon, for example — are bordered by great circles, but we see them as elliptical arcs. Tilt a glass of water (it doesn't matter if the glass has cylindrical or conical sides) and the surface of the liquid acquires an elliptical outline.

A ball resting on a table top [*see Fig. 92*] casts an elliptical shadow that is a cross section of a cone of light in which the ball fits snugly. The ball rests precisely on one focus of the shadow. If we imagine a larger sphere that is tangent to the surface from beneath and fits snugly into the same cone, the larger sphere will touch the shadow at the other focus. These two spheres provide the following famous and magnificent proof (by G. P. Dandelin, a 19th-century Belgian mathematician) that the conic section is indeed an ellipse.

Point A is any point on the ellipse. Draw a line [*shown in color in the illustration*] that passes through A and the apex of the cone. This line will be tangent to the spheres at points D and E. Draw a line from A to point B, where the small sphere touches the shadow, and a similar line from A to C, where the large sphere

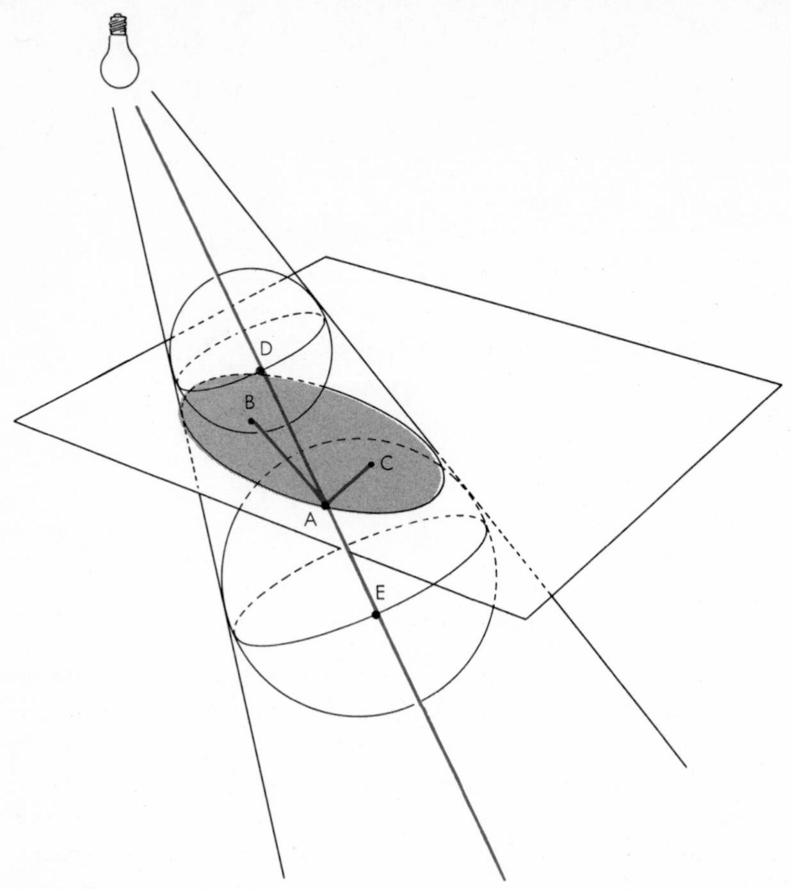

FIG. 92
By means of larger sphere it can be shown that shadow of smaller sphere is an ellipse.

touches the shadow. AB is equal to AD because both lines are tangents to a sphere from the same fixed point. AE equals AC for the same reason. Adding equals to equals:

$$AD + AE = AB + AC$$

Now AD + AE is the same as the straight line DE. Because of the symmetry of cone and spheres, this line has a constant length regardless of where point A is chosen on the ellipse. If the sum of AD and AE is constant, then the above equation makes the sum of AB and AC a constant also. Since AB and AC are the distances of point A from two fixed points, the locus of A must be an ellipse with B and C as its two foci.

In physics the ellipse turns up most often as the path of an object moving in a closed orbit under the influence of a central force that varies inversely with the square of the distance. Planets and satellites, for example, have elliptical orbits with the center of gravity of the parent body at one of the foci. When Kepler first announced his great discovery that planets move in ellipses, it ran so counter to the general belief that God would not permit the paths of heavenly bodies to be less perfect than circles that Kepler found it necessary to apologize. He spoke of his ellipses as dung that he had been forced to introduce in order to sweep from astronomy the larger amount of dung that had accumulated around attempts to preserve circular orbits. Kepler himself never discovered why the orbits were elliptical; it remained for Newton to deduce this from the nature of gravity. Even the great Galileo to his dying day refused to believe, in the face of mounting evidence, that the orbits were not circular.

An important reflection property of the ellipse is made clear in Figure 93. Draw a straight line that is tangent to the ellipse at any point. Lines from that point to the foci make equal angles with the tangent. If we think of the ellipse as a vertical strip of metal on a flat surface, then any body or wave pulse, moving in a straight line from one focus, will strike the boundary and rebound directly toward the other focus. Moreover, if the body or wave is moving toward the boundary at a uniform rate, regardless of the direction it takes when it leaves one focus, it is sure to rebound to the other focus after the same time interval (since the two distances have a constant sum). Imagine a shallow elliptical tank filled with water. We start a circular wave pulse by dipping a finger into the water at one focus of the ellipse. A moment later there is a convergence of circular waves at the other focus.

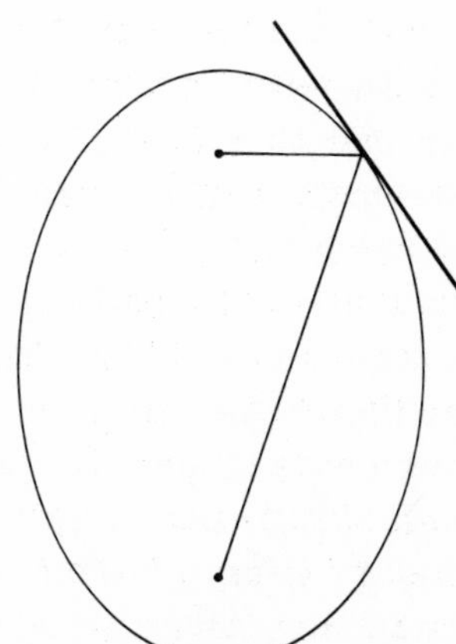

FIG. 93
Tangent makes equal angles with two lines.

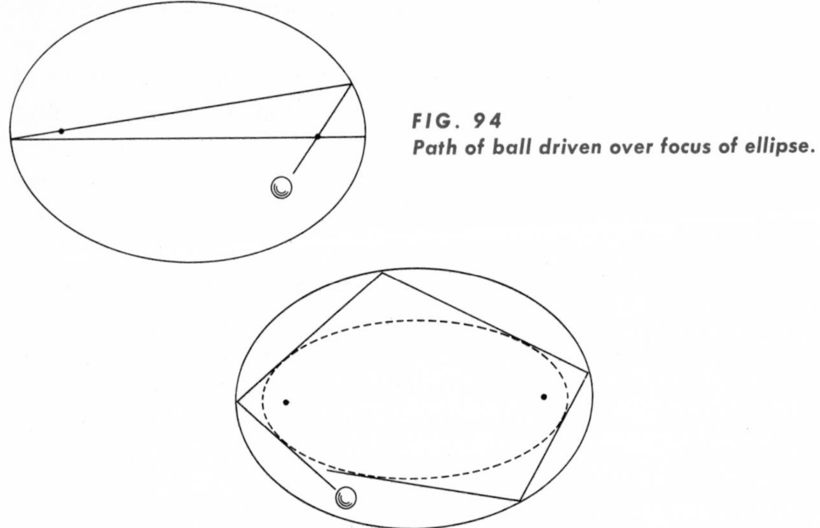

FIG. 94
Path of ball driven over focus of ellipse.

Path of ball that does not go between foci.

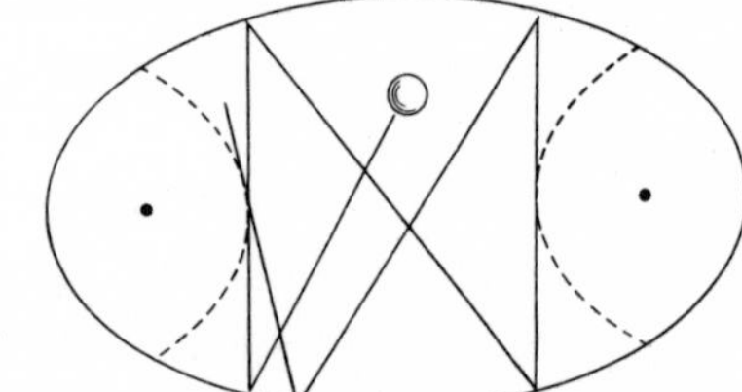

Path of ball that does pass between foci.

Lewis Carroll invented and published a pamphlet about a circular billiard table. I know of no serious proposal for an elliptical billiard table, but Hugo Steinhaus (in his book *Mathematical Snapshots*, recently reissued in a revised edition by the Oxford University Press) gives a surprising threefold analysis of how a ball on such a table would behave. Placed at one focus and shot (without English) in any direction, the ball will rebound and pass over the other focus. Assuming that there is no friction to retard the motion of the ball, it continues to pass over a focus with each rebound [*see top illustration of Fig. 94*]. However, after only a few trips the path becomes indistinguishable from the ellipse's major axis. If the ball is not placed on a focus, then driven so that it does not pass between the foci, it continues forever along paths tangent to a smaller ellipse with the same foci [*see middle illustration of Fig. 94*]. If the ball is driven between the foci [*see bottom illustration of Fig. 94*], it travels endlessly along paths that never get closer to the foci than a hyperbola with the same foci.

In *The Mikado* there are lines about a billiard player forced to play

> *On a cloth untrue*
> *With a twisted cue,*
> *And elliptical billiard balls!*

In *A Portrait of the Artist as a Young Man* James Joyce has a teacher quote these lines, then explain that by "elliptical" W. S. Gilbert really meant "ellipsoidal." What is an ellipsoid? There are three principal types. An ellipsoid of rotation, more properly called a spheroid, is the surface of a solid obtained by rotating an ellipse around either axis. If the rotation is around the minor axis, it generates an oblate spheroid, which is flattened at the poles like the earth. Rotation around the major axis generates the football-shaped prolate spheroid. Imagine a prolate spheroid surface that is a mirror on the inside. If a candle is lighted at one focus, a piece of paper at the other focus will burst into flames.

Whisper chambers are rooms with spheroidal ceilings. Faint sounds originating at one focus can be heard clearly at the other focus. In the U. S. the best-known whispering gallery is in Statuary Hall of the Capitol in Washington, D.C. (No guided tour is complete without a demonstration.) A smaller but excellent whisper chamber is a square area just outside the entrance to the Oyster Bar on the lower level of New York's Grand Central Station. Two people standing in diagonally opposite corners, facing the wall, can hear each other distinctly even when the square area bustles with activity.

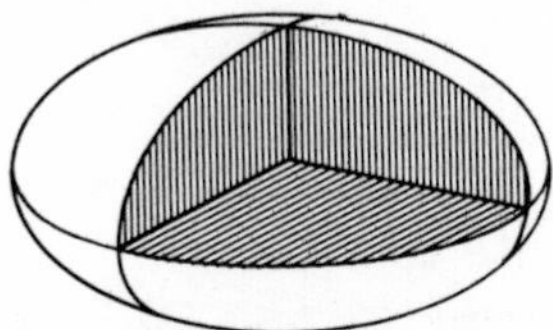

FIG. 95
Each section of ellipsoid is elliptical.

Both the oblate and prolate spheroids have circular cross sections if sliced by planes perpendicular to one of the three coordinate axes, elliptical cross sections if sliced by planes perpendicular to the other two axes. When all three axes are unequal in length, and sections perpendicular to each are ellipses, the shape is a true ellipsoid [*see Fig. 95*]. This is the shape that pebbles on a beach tend to assume after long periods of being jostled by the waves.

Elliptical "brainteasers" are rare. Here are two easy ones.

1. Prove that no regular polygon having more sides than a square can be drawn on a noncircular ellipse so that each corner is on the perimeter of the ellipse.

2. In the paper-folding method of constructing an ellipse, explained earlier, the center of the circle and the spot on the circle are the two foci. Prove that the curve outlined by the creases is really an ellipse.

ADDENDUM

HENRY DUDENEY, in problem 126 of *Modern Puzzles,* explains the string-and-pins method of drawing an ellipse, then asks how one can use this method for drawing an ellipse with given major and minor axes. The method is simple:

First draw the two axes. The problem now is to find the two foci, A and B, of an ellipse with these axes. Let C be an end of the minor axis. Points A and B are symmetrically located on the major axis at spots such that AC and CB each equals half the length of the major axis. It is easy to prove that a loop of string with a length equal to the perimeter of triangle ABC will now serve to draw the desired ellipse.

Elliptical pool tables actually went on sale in the United States in 1964. A full-page advertisement in *The New York Times* (July 1, 1964) announced that on the following day the game would be introduced at Stern's department store by Broadway stars Joanne Woodward and Paul Newman. Elliptipool, as it is called, is the patented invention of Arthur Frigo, Torrington, Connecticut, then a graduate student at Union College in Schenectady. Because the table's one pocket is at one of the foci, a variety of weird cushion shots can be made with ease.

The eleventh edition of *Encyclopaedia Britannica* in its article on billiards has a footnote that reads: "In 1907 an oval table was introduced in England by way of a change." Neither this table nor Lewis Carroll's circular table had a pocket, however. A design patent (198,571) was issued in July 1964 to Edwin E. Robinson, Pacifica, California, for a circular pool table with four pockets.

ANSWERS

1. No regular polygon with more sides than a square can be inscribed in an ellipse for this reason: The corners of all regular polygons lie on a circle. A circle cannot intersect an ellipse at more than four points. Therefore, no regular polygon with more than four corners can be placed with all its corners on an ellipse. This problem was contributed by M. S. Klamkin to *Mathematics Magazine* for September-October 1960.

2. The proof that the paper-folding method of constructing an ellipse actually does produce an ellipse is as follows. Let point A in Figure 96 be any point on a paper circle that is not the circle's center (O). The paper is folded so that any point (B) on the circumference falls on A. This creases the paper along XY. Because XY is the perpendicular bisector of AB, BC must equal AC. Clearly OC + AC = OC + CB. OC + CB is the circle's radius, which cannot vary, therefore OC + AC must also be constant. Since OC + AC is the sum of the distances of point C from two fixed points A and O, the locus of C (as point B moves around the circumference) must be an ellipse with A and O as the two foci.

The crease XY is tangent to the ellipse at point C because it makes equal angles with the lines joining C to the foci. This is easily established by noting that angle XCA equals angle XCB, which in turn equals angle YCO. Since the creases are always tangent to the ellipse, the ellipse becomes the envelope of the infinite set of creases that can be produced by repeated folding of the paper. This proof is taken from Donovan A. Johnson's booklet *Paper Folding for the Mathematics Class,* published in 1957 by the National Council of Teachers of Mathematics.

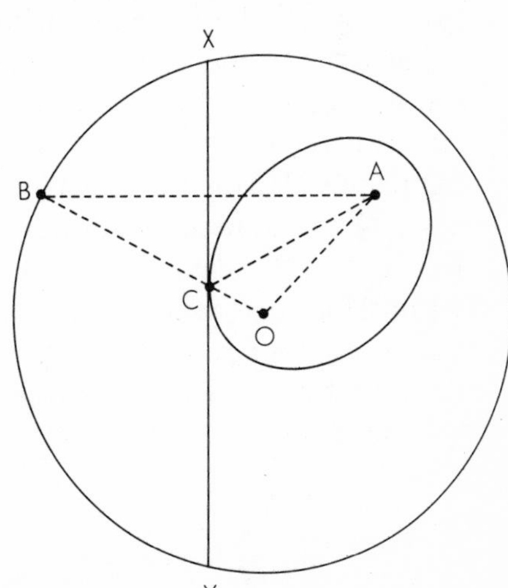

FIG. 96
Answer to the paper-folding problem.

CHAPTER SIXTEEN

□

The 24 Color Squares and the 30 Color Cubes

IN THE U.S. a standard set of dominoes consists of 28 oblong black tiles, each divided into two squares that are either blank or marked with white spots. No two tiles are alike, and together they represent the 28 possible ways in which numbers from 0 to 6 can be combined two at a time. The tiles can be regarded as line segments that are placed end to end to form linear chains; in this sense all domino games are strictly one-dimensional. When the domino concept is extended to two- and three-dimensional pieces, all sorts of colorful and little-known recreations arise. Percy Alexander MacMahon, a British authority on combinatorial analysis, devoted considerable thought to these superdominoes, and it is from his book *New Mathematical Pastimes,* published in 1921, that much of the following material is taken.

For a two-dimensional domino the equilateral triangle, square and hexagon are the most convenient shapes because in each case identical regular polygons can be fitted together to cover a plane completely. If squares are used and their edges are labeled in all possible ways with n different symbols, a set of $\frac{1}{4}n(n + 1)(n^2 - n + 2)$ squares can be formed. Figure 97 shows the full set of 24 square dominoes that results when $n = 3$. If the reader constructs such a set from cardboard he will have the equipment for a first-rate puzzle. Colors are easier to work with than symbols, so it is suggested that colors be used instead of symbols. The

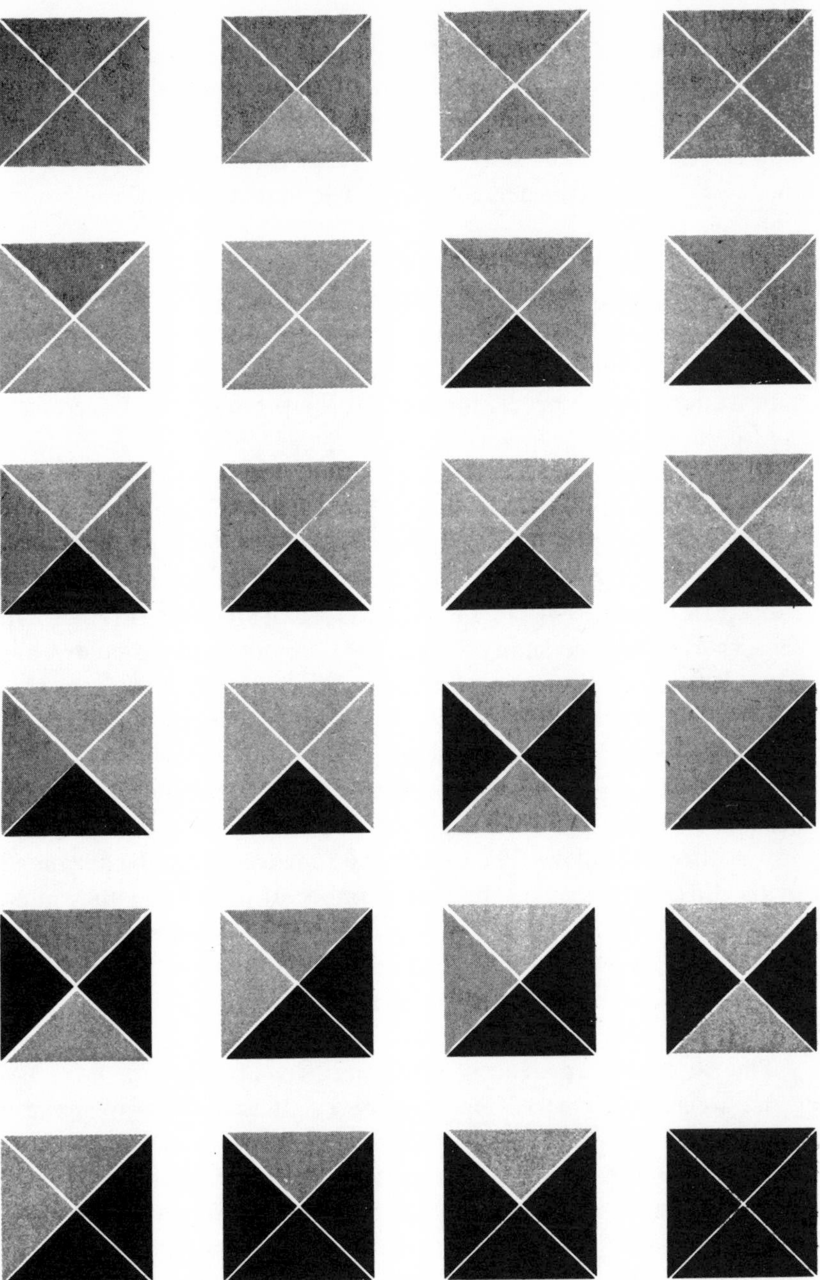

FIG. 97
A set of square dominoes using three colors.

problem is to fit together all 24 squares into a four-by-six rectangle, with two provisos: (1) each pair of touching edges must be the same color; (2) the border of the rectangle, all the way around, must be the same color. It is assumed that the cardboard squares are colored on only one side. Any color may be picked for the border, and with each choice, a large number of different solutions are possible.

The four-by-six rectangle is the only one that can be formed under the given restrictions. A two-by-twelve is obviously impossible because it would require that each square have a triangle of the border color. Can the reader, simply by looking over the 24 color squares in Figure 97, prove that the three-by-eight rectangle is also impossible?

In three dimensions, cubes are the only regular solids that will press together to fill a three-dimensional space completely; for this reason they are the most satisfactory shapes for 3-D dominoes. If two colors are used for the faces, no more than ten different cubes can be painted — a number too small to be of interest. On the other hand, too many cubes (57) result if three colors are used. With six colors the number jumps to 2,226, but from this set we can pick a subset of 30 that is ideal for our purposes. It consists of cubes that bear all six colors on their six faces.

It is easy to see that 30 is the maximum number. There must be, say, a red face on each cube. Opposite this face can be any one of five different colors. The remaining four colors can be arranged in six different ways, so the total number of different cubes must be $5 \times 6 = 30$. (Two cubes are considered different if it is impossible to place them side by side in such a way that all corresponding faces match.) Figure 98 shows the 30 cubes in "unfolded" form.

The 30 cubes, apparently discovered by MacMahon, have become a classic of recreational geometry. It is a chore to make a set, but the effort brings rich rewards. A set of neatly painted cubes is an endlessly fascinating family toy; it requires no batteries and is unlikely to wear out for decades. Wooden or plastic blocks, preferably with smooth sides, can be bought at most toy counters or obtained from a friend with access to a buzz saw. An alternative to painting is to paste squares of colored paper on the cubes.

FIG. 98
The 30 color cubes unfolded.

For an introductory exercise, pick any one of the 30 cubes. Now find a second cube that can be placed face to face with the first one so that the touching faces match, the end faces are a second color and the other four colors are on the four sides, each side a solid color. It is always possible to do this. Since the two cubes are mirror images of each other, this means that every cube, like every fundamental particle of matter, has its anticube.

(In searching for a certain type of cube, much time can be saved by lining them up in rows and turning an entire row at once by applying pressure at the ends. For example, suppose you are looking for cubes with red and blue on opposite sides. Arrange a group of cubes in a row with red on top, give the row two quarter-turns and take out all cubes that now show blue on top. Or suppose you wish to work with cubes on which blue, yellow and green touch at the same corner. Arrange a row with all blue on top, invert it and discard the greens and yellows. Turn the remaining cubes to show green on top, invert them and discard the blues and yellows. The cubes left will be of the desired type.)

It is not possible to form a straight chain of more than two cubes and have each of the four sides a solid color, but a row of six is easily made that has all six colors on each side. A pretty problem is to do this with all touching faces matching and the two end faces also matching.

Now for a more difficult puzzle. Choose any cube and place it to one side. From the remaining 29 select eight that can be formed into a two-by-two-by-two cube that is an exact model of the chosen one except twice as high. In addition, each pair of touching faces must match. (MacMahon credits the discovery that this can always be done, regardless of which cube is chosen, to his friend Colonel Julian R. Jocelyn.)

Only one set of eight cubes will do the trick, and they are not easy to find without a systematic procedure. The following is perhaps the best. Note the three pairs of opposite faces on the prototype, then eliminate from the 29 cubes all those that have a pair of opposite faces corresponding to any of the three pairs on the prototype. Sixteen cubes will remain. Turn the prototype so that one of its top corners points toward you and only the three faces meeting at that corner are visible. Among the sixteen cubes you will find two that can be placed so that the same three faces are in the same position as the three on the prototype. Put these two aside. Turn the cube so that another top corner points toward you and find the two cubes that match this corner. The eight cubes selected in this way — two for each top corner of the prototype — are the cubes required. It is now a simple task to build the model.

Actually there are two essentially different ways to build the model with these eight cubes. L. Vosburgh Lyons, a Manhattan neuropsychiatrist, devised the ingenious procedure, depicted in Figure 99, by which any model can be changed to its second form. The two models are related in remarkable ways. The 24 outside faces of each model are the 24 inner faces of the other, and when the two models are similarly oriented, each cube in one is diagonally opposite its location in the other.

Lyons has discovered that after a model has been built it is always possible to select a new prototype from the remaining 21 cubes, then build a two-by-two-by-two model of the new prototype with eight of the remaining twenty. Few succeed in doing this unless they are tipped off to the fact that the new prototype must be a mirror image of the first one. The eight cubes needed for the model are the eight rejected from the sixteen in the last step of the procedure by which the cubes were chosen for the first prototype.

Many other color-cube construction puzzles have been proposed. The following two-by-two-by-two models, all possible, are taken from *Das Spiel der 30 Bunten Würfel,* by Ferdinand Winter, a book on the color cubes, published in Leipzig in 1934. In all these models the cubes must obey the domino rule of having touching faces of like color.

1. One color on left and right faces, second color on front and back, third color on top, fourth on bottom.

2. One color on two opposite faces, different colors on the other four.

3. One color on left and right faces, second color on front and back, the remaining four colors on top (each square a different color), and the same four colors on bottom.

4. Each face is four-colored, the same four colors on every face.

Apparently it is not possible to build a two-by-two-by-two cube with one color on front and back, a second color on left and right,

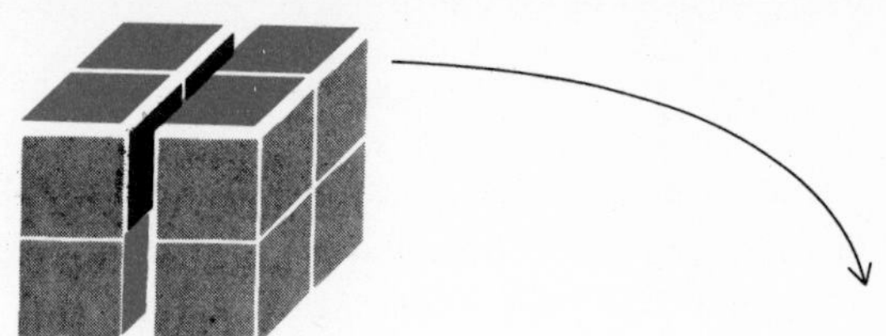

1. MODEL SHOWN HERE HAS RED ON TOP, BLACK ON BOTTOM. TURN SO INTERIOR RED AND BLACK FACES ARE IN POSITION SHOWN. MOVE TOP HALF OF MODEL TO RIGHT.

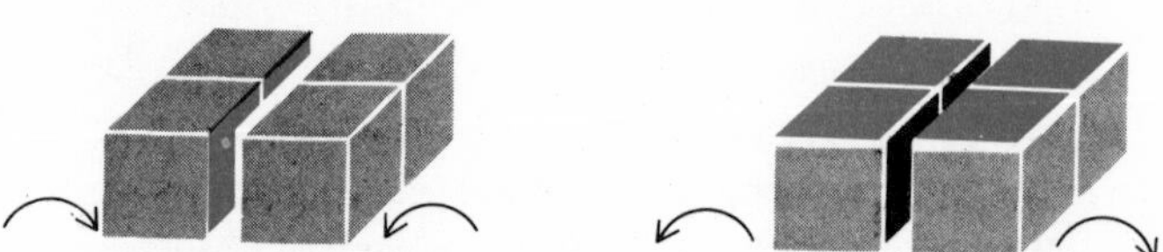

2. GIVE EACH COLUMN A QUARTER-TURN IN DIRECTION SHOWN BY ARROW, TO FORM R FACE ON BOTTOM OF LEFT SQUARE, BLACK FACE ON TOP OF RIGHT SQUARE.

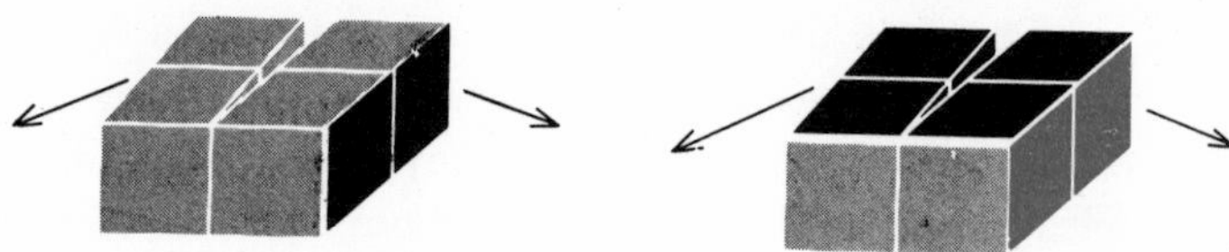

3. UNFOLD EACH SQUARE, BRINGING ENDS "A" TOGETHER TO FORM TWO ROWS.

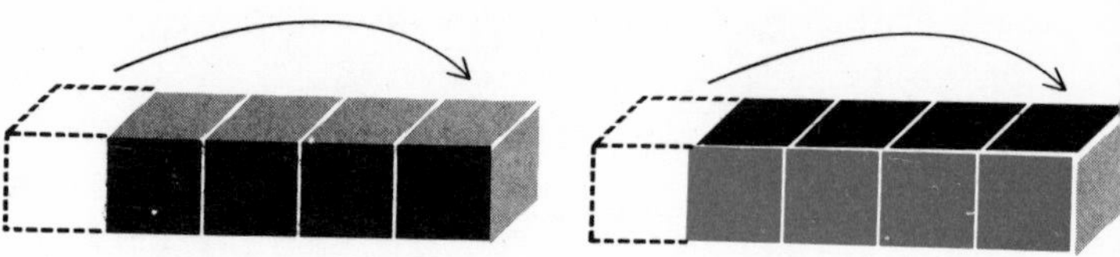

4. MOVE A CUBE FROM LEFT TO RIGHT END OF EACH ROW.

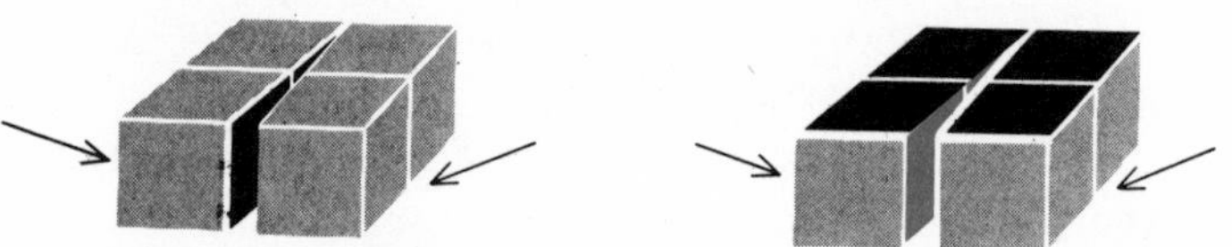

5. FOLD EACH ROW IN HALF, BRINGING BLACK FACES TOGETHER ON LEFT, RED FACES TOGETHER ON RIGHT.

FIG. 99
The Lyons method of transforming a model to its second form.

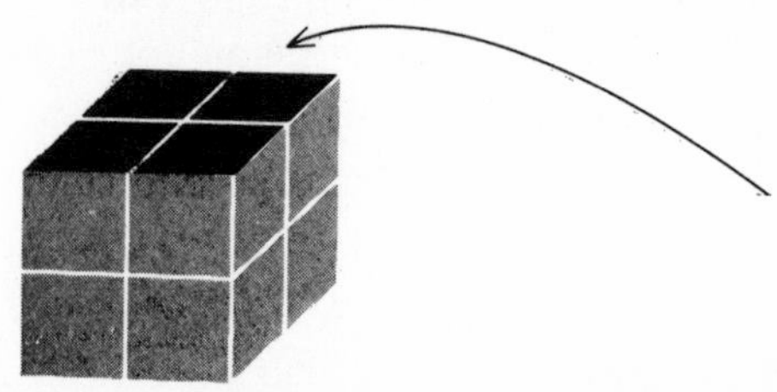

6. PUT RIGHT SQUARE ON TOP OF LEFT. SECOND FORM OF MODEL IS NOW COMPLETE.

third on top and bottom, and all touching faces matching. It is possible to build a three-by-three-by-three cube with each face a different color, but not without violating the domino rule about touching faces.

Games of the domino type can be played with any species of two- or three-dimensional domino; in fact, Parker Brothers still sells a pleasant game called Contack (first brought out by them in 1939), which is played with equilateral-triangle tiles. Of several games that have been proposed for the color cubes, a game called Color Tower seems the best.

Two players sit opposite each other. Each has in front of him a screen that is easily made by taking a long strip of cardboard about ten inches wide and folding the ends to make it stand upright. The cubes are put into a container in which they cannot be seen but from which they can be taken one at a time. A paper bag will do, or a cardboard box with a hole in the top.

Each player draws seven cubes from the container and places them behind his screen, where they are hidden from his opponent. The first player opens the game by placing a cube in the middle of the table. (The privilege of opening can be decided by rolling a cube after a player has named three colors. If one of the three comes up, he plays first.) The second player then places a cube against the side of the first one, touching faces matching. Players alternate turns, each adding one cube to the structure, and in this way build a tower that rests on a square base of four cubes. A player's object is to get rid of all his cubes.

The rules are as follows:

1. Each tier of four cubes must be completed before starting the next tier.

2. A cube may be placed in any open spot on a tier, provided that it meets two conditions: all touching faces must match, and it must not make impossible any remaining play on the tier. In Figure 100, for example, cube A would be illegally played if any of its faces met at right angles with an exposed face of the same color.

3. If a player cannot play any of his cubes, he must draw one from the container. If the drawn cube is playable, he may play it if he wishes. If he cannot or does not wish to play it, he awaits his next turn.

FIG. 100
The game of Color Tower.

4. If for strategic reasons a player wishes to pass up his turn, he may do so at any time, but he must draw a cube from the container.

5. The game ends when one player is rid of all his cubes. He scores 3 points for winning, plus the number of cubes that remain in his opponent's hand.

6. If all cubes are drawn from the container, turns alternate until one player is unable or unwilling to play. The other player then plays until his opponent is able or willing to continue. If both are unable or unwilling to play, the game ends and the person with the smallest hand is the winner. He scores the difference between hands.

7. The goal of a set of games can be any agreed-upon number of points. If played as a gambling game, the winner collects after each game an amount equal to his score.

Various strategies occur to anyone who plays Color Tower for a while. Suppose your opponent has just started a new tier. You have two cubes left. It would be unwise to play diagonally opposite his cube in such a way as to make your last cube unplayable in either of the remaining three-face plays. It may be necessary to play alongside his cube to keep open the possibility of going out on your next move. The discovery of such strategies makes the learning of Color Tower a stimulating experience and leads to a skill in play that greatly increases one's probability of winning.

If any reader has suggestions for improving Color Tower I would enjoy hearing about them, as well as about any other games or unusual new puzzles with the cubes. The 30 color cubes have been around for almost 50 years, but they probably contain many more surprises.

ADDENDUM

WHEN I explained MacMahon's puzzle with the 24 color squares, I made the blunder (I had misinterpreted one of MacMahon's comments) of saying that it had only one solution. This proved to be the greatest understatement ever made in the column. First, I heard from about fifty readers who sent more than one pattern. Thomas O'Beirne devoted his column, "Puzzles and Paradoxes," in *The New Scientist* (February 2, 1961) to the puzzle and showed how dozens of solutions could be obtained.

In Buenos Aires the problem caught the interest of Federico Fink. He and his friends found hundreds of distinct solutions (rotations and reflections are not, of course, counted as different), and over the months his list grew into the thousands. On November 20, 1963, he wrote to say that he estimated the total number of different patterns to be 12,224.

The matter was settled early in 1964. Fink suggested to Gary Feldman, at Stanford University's Computation Center, that he write a computer program for the puzzle. Feldman did. The center's B5000, using a program written in ALGOL and running about 40 hours, produced a complete list of all possible patterns. There are 12,261. Fink missed by only 37, a truly amazing prediction. Feldman's account of his program, "Documentation of the MacMahon Squares Problem," a Stanford Artificial Intelligence Project Memo No. 12, was issued as an eight-page typescript by the Computation Center on January 16, 1964.

It would take many pages to summarize the main results of Fink's analysis of the 12,261 solutions. None of the patterns, regretfully, exhibits bilateral symmetry. The maximum number of "diamonds" (single-color squares formed by two right triangles) that can come together to form a polyomino of one color is twelve.

Figure 101, left, shows such a pattern; the order-12 polyomino has bilateral symmetry and resembles a lobster. The minimum number of "isolated diamonds" (diamonds completely surrounded by other colors) is three. Figure 101, middle, is a pattern in which each of the three isolated diamonds is a different color. The maximum number of isolated diamonds is thirteen, as exemplified by Figure 101, right.

Note that all three patterns show a "bridge" of three isolated diamonds, of the border color, that join right and left borders. O'Beirne, in his *New Scientist* column, proved that every solution must have such a bridge. The bridge's position, together with the other spots of border color, provide a convenient classification of twenty different species of solution. (O'Beirne listed eighteen, but Fink later found two more.)

Many recreations involving color cubes await exploration. For instance, from the set of 57 cubes with one, two and three colors, one can pick the 27 that bear no more than two different colors on any one cube. Since 27 cubes form one three-by-three-by-three, there may be some good construction problems here. Or one could work with the subset of 30 that have three different colors on each cube. Some of the constructions not possible with the 30 six-color cubes may be possible with these 30 three-color cubes. Can an all-red cube be made, for example, under the usual restriction that touching faces be the same color?

FIG. 101
Three of the 12,261 solutions to the color-square problem: the lobster (left), *three isolated diamonds of different colors* (middle), *and 13 isolated diamonds* (right).

MacMahon, who presumably invented the 30-color-cube recreation, was a major in England's Royal Artillery who taught mathematics at the Royal Military Academy. He is best known for his *Introduction to Combinatory Analysis* and his article on the topic in the eleventh edition of *Encyclopaedia Britannica*. He died in 1928. Thomas O'Beirne informs me that a set of eight color cubes, to be formed into one larger cube according to certain provisos, was once sold in England as the Mayblox puzzle, with credit on the box's cover to MacMahon as the inventor.

A popular puzzle often found on sale in different countries, under various trade names, consists of four cubes, each colored with four different colors. The puzzle is to arrange them in one row so that all four colors (in any order) appear on each side of the four-by-one square prism. Sometimes symbols, such as the four card suits or advertising pictures of products, appear instead of colors. For descriptions of such puzzles, see R. M. Abraham, *Diversions and Pastimes*, Dover, 1964, page 100; and Anthony Filipiak, *100 Puzzles*, A. S. Barnes, 1942, page 108. A thorough analysis of puzzles of this type will be found in Chapter 7, "Cubism and Colour Arrangements," of O'Beirne's *Puzzles and Paradoxes*, Oxford University Press, 1965.

ANSWERS

THREE SAMPLE solutions for MacMahon's color-square puzzle were given in the addendum. Solutions for the color-cube problems are left for readers to discover.

To prove that the three-by-eight rectangle cannot be formed with the 24 color squares, to meet the imposed conditions, first select any four squares, with adjacent triangles of the same color, to go in the four corners. Exactly fourteen squares, bearing the same color, remain; just enough for the fourteen remaining border cells of the rectangle. At least three of them, however, will have the border color on opposite sides, calling for three internal squares bearing the same color. But there are *no* more squares with this color; all have been used for the border. The three-by-eight, therefore, is an impossible rectangle.

CHAPTER SEVENTEEN

□

H. S. M. Coxeter

MOST PROFESSIONAL mathematicians enjoy an occasional romp in the playground of mathematics in much the same way that they enjoy an occasional game of chess; it is a form of relaxation that they avoid taking too seriously. On the other hand, many creative, well-informed puzzlists have only the most elementary knowledge of mathematics. H. S. M. Coxeter, professor of mathematics at the University of Toronto, is one of those rare individuals who are eminent as mathematicians and as authorities on the not-so-serious side of their profession.

Harold Scott Macdonald Coxeter was born in London in 1907 and received his mathematical training at Trinity College, Cambridge. On the serious side he is the author of *Non-Euclidean Geometry* (1942), *Regular Polytopes* (1948) and *The Real Projective Plane* (1955). On the lighter side he has edited and brought up to date W. W. Rouse Ball's classic work *Mathematical Recreations and Essays*, and has contributed dozens of articles on recreational topics to various journals. In 1961 John Wiley & Sons published his *Introduction to Geometry*, a book that is the topic of this chapter.

There are many ways in which Coxeter's book is remarkable. Above all, it has an extraordinary range. It sweeps through every branch of geometry, including such topics as non-Euclidean geometry, crystallography, groups, lattices, geodesics, vectors, projective geometry, affine geometry and topology — topics not always found in introductory texts. The writing style is clear, crisp and for the most part technical. It calls for slow, careful reading but has the merit of permitting a vast quantity of material to be compressed between its covers. The book is touched throughout with the author's sense of humor, his keen eye for mathematical beauty and his enthusiasm for play. Most of its sections open with apt literary quotations, many from Lewis Carroll, and close with exercises that are often new and stimulating puzzles. A number of sections deal entirely with problems and topics of high recreational interest, some of which have been discussed, on a more elementary level, in this and the two previous volumes of this series: the golden ratio, regular solids, topological curiosities, map coloring, the packing of spheres and so on.

Amusing bits of off-trail information dot the text. How many readers know, for example, that in 1957 the B. F. Goodrich Company patented the Moebius strip? Its patent, No. 2,784,834, covers a rubber belt that is attached to two wheels and is used for conveying hot or abrasive substances. When the belt is given the familiar half-twist, it wears equally on both sides — or rather on its single side.

And how many readers know that at the University of Göttingen there is a large box containing a manuscript showing how to construct, with compass and straightedge only, a regular polygon of 65,537 sides? A polygon with a prime number of sides can be constructed in the classical manner only if the number is a special type of prime called a Fermat prime: a prime that can be expressed as $2^{(2^n)} + 1$. Only five such primes are known: 3, 5, 17, 257 and 65,537. The poor fellow who succeeded in constructing the 65,537-gon, Coxeter tells us, spent ten years on the task. No one knows whether there is a prime-sided polygon larger than this that is in principle constructible with compass and straightedge. If there is such a polygon, its actual construction would be out of the question, since the number of sides would be astronomical.

It might be supposed that the lowly triangle, studied so thoroughly by the ancients, would contain few new surprises. Yet many remarkable theorems about the triangle — theorems that Euclid could easily have discovered but didn't — have been found only in recent times. One outstanding example, discussed by

Coxeter, is Morley's theorem. It was first discovered about 1899 by Frank Morley, professor of mathematics at Johns Hopkins University and father of the writer Christopher Morley. It spread rapidly through the mathematical world in the form of gossip, Coxeter writes, but no proof for it was published until 1914. When Paul and Percival Goodman, in Chapter 5 of their wonderful little book *Communitas,* speak of human goods that are not consumed while being enjoyed, it is Morley's beautiful theorem that provides a happy illustration.

Morley's theorem is illustrated in Figure 102. A triangle of any shape is drawn, and its three angles are trisected. The trisecting lines always meet at the vertices of an equilateral triangle. It is the appearance of that small equilateral triangle, known as the Morley triangle, that is so totally unexpected. Professor Morley wrote several textbooks and did important work in many fields, but it is this theorem that has earned him his immortality. Why was it not discovered earlier? Coxeter thinks that perhaps mathematicians, knowing the angle could not be trisected within the classical limitations, tended to shy away from theorems involving angle trisections.

Another triangle theorem that has achieved widespread notoriety in this century is illustrated in Figure 103. If the internal bisectors of the two base angles of a triangle are equal, it seems intuitively obvious that the triangle must be isosceles. But can you prove it? No problem in elementary geometry is more insidiously deceptive. Its converse—the bisectors of the base angles of an isosceles triangle are equal — goes back to Euclid and is easy to prove. This one *looks* as if a proof would be just as easy, when in fact it is extremely difficult. Every few months someone

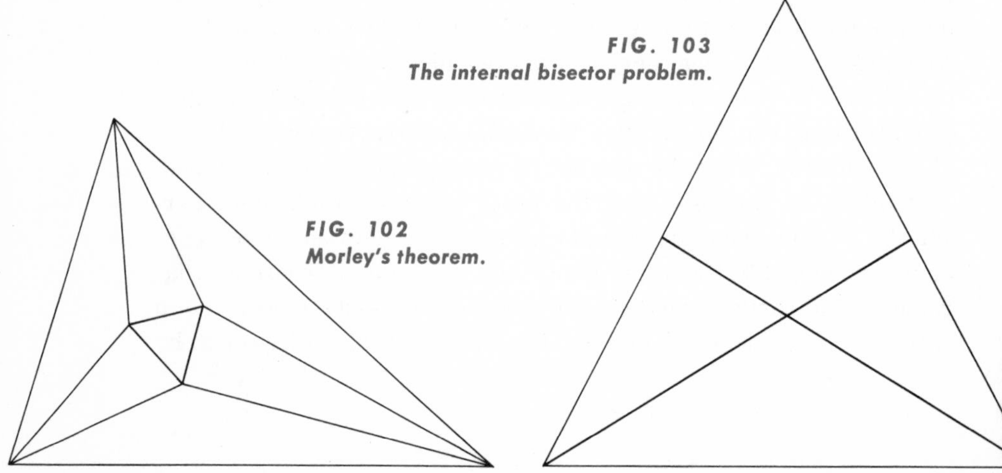

FIG. 103
The internal bisector problem.

FIG. 102
Morley's theorem.

sends me a plea for a proof of this problem. I usually reply by citing an article by Archibald Henderson that appeared in the *Journal of the Elisha Mitchell Scientific Society* for December 1937. Henderson calls his paper, almost 40 pages long, "an essay on the internal bisector problem to end all essays on the internal bisector problem." He points out that many published proofs, some by famous mathematicians, are faulty; then he gives ten valid proofs, all long and involved. It is a pleasant shock to find in Coxeter's book a new proof, so simple that all he need do is devote four lines to a hint from which the proof is easily derived.

Now and then, when someone discovers an elegant new theorem, he is moved to record it in verse. An amusing modern instance is "The Kiss Precise," a poem by the distinguished chemist Frederick Soddy, who coined the word "isotope." If three circles of any size are placed so that each touches the other two, it is always possible to draw a fourth circle that touches the other three. Usually there are two ways to draw a fourth circle; sometimes one is a large circle enclosing the other three. In Figure 104, for instance, the two possible fourth circles are shown as

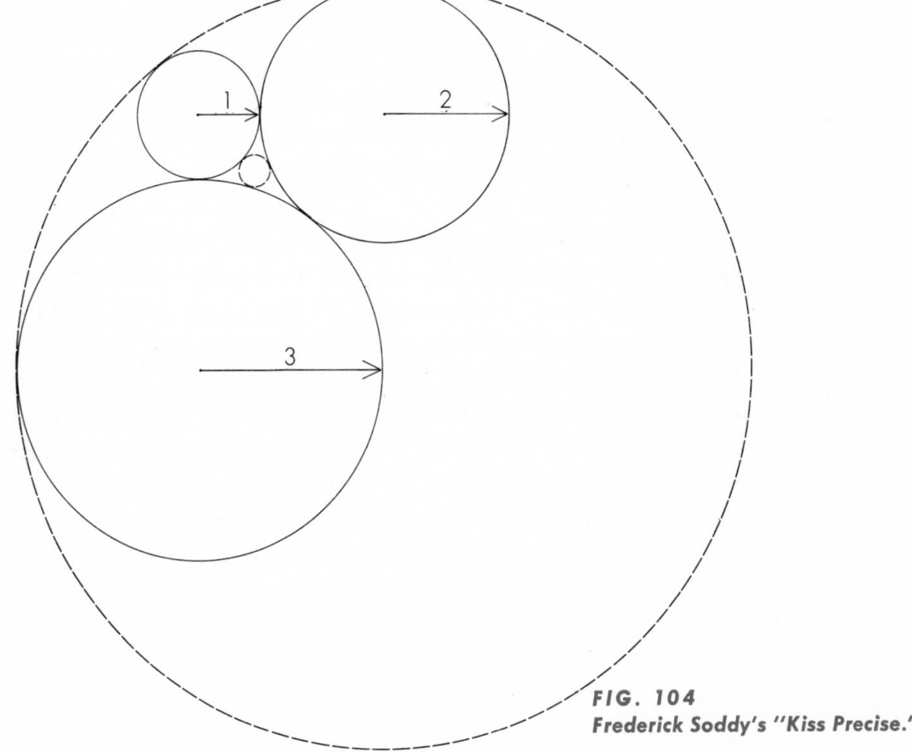

FIG. 104
Frederick Soddy's "Kiss Precise."

broken lines. How are four mutually tangent circles related to each other in size? Soddy, as the result of a procedure that he later confessed he never really understood, chanced upon the following beautifully symmetrical formula, in which a, b, c and d are the reciprocals of the four radii:

$$a^2 + b^2 + c^2 + d^2 = \tfrac{1}{2}^2 \, (a + b + c + d)^2$$

The reciprocal of a number n is simply $1/n$, and the reciprocal of any fraction is obtained by turning the fraction upside down. The reciprocal of a radius is the measure of a circle's curvature. A concave curvature, such as that of a circle enclosing the other three, is considered a negative curvature and is handled as a negative number. In his poem Soddy uses the term "bend" for curvature. Coxeter quotes the second stanza of the poem as follows:

Four circles to the kissing come,
The smaller are the benter.
The bend is just the inverse of
The distance from the centre.
Though their intrigue left Euclid dumb
There's now no need for rule of thumb.
Since zero bend's a dead straight line
And concave bends have minus sign,
The sum of the squares of all four bends
Is half the square of their sum.

Soddy's formula is a great timesaver for puzzlists; problems involving kissing circles, often found in puzzle books, are tough to crack without it. For example, if the three solid circles in Figure 104 have radii of one, two and three inches, what are the radii of the broken circles? This can be answered by drawing a large number of right triangles and doggedly applying the Pythagorean theorem, but Soddy's formula gives a simple quadratic equation with two roots that are the reciprocals of the two radii sought. The positive root gives the small broken circle a curvature of 23/6 and a radius of 6/23 inches; the negative root gives the large broken circle a negative curvature of $-1/6$ and a six-inch radius.

Readers who care to test the formula's power on another problem can consider this situation. A straight line is drawn on a plane. Two kissing spheres, one with a radius of four inches, the other with a radius of nine inches, stand on the line. What is the radius of the largest sphere that can be placed on the same line so that it kisses the other two? Instead of Soddy's formula one can use the following equivalent expression, supplied by Coxeter, which makes the computation much easier. Given the three reciprocals, a, b and c, the fourth reciprocal is:

$$a + b + c \pm 2\sqrt{ab + bc + ac}$$

From an artist's point of view, some of the most striking pictures in Coxeter's richly illustrated volume accompany his discussions of symmetry and the role played by group theory in the construction of repeated patterns such as are commonly seen in wallpaper, tile flooring, carpeting and so on. "A mathematician, like a painter or a poet, is a maker of patterns," wrote the English mathematician G. H. Hardy in a famous passage quoted by Coxeter. "If his patterns are more permanent than theirs, it is because they are made with *ideas*." When polygons are fitted together to cover a plane with no interstices or overlapping, the pattern is called a tessellation. A regular tessellation is one made up entirely of regular polygons, all exactly alike and meeting corner to corner (that is, no corner of one touches the side of another). There are only three such tessellations: a network of equilateral triangles, the checkerboard pattern of squares, and the hexagonal pattern of the honeycomb, chicken wire and bathroom tiling. The squares and triangles can also be made to fill the plane without placing them corner to corner, but this cannot be done with the hexagons.

"Semiregular" tessellations are those in which two or more kinds of regular polygons are fitted together corner to corner in such a way that the same polygons, in the same cyclic order, surround every vertex. There are precisely eight of these tessellations, made up of different combinations of triangles, squares, hexagons, octagons and dodecagons [*see Fig. 105*]. All of them would, and some do, make excellent linoleum patterns. All are unchanged by mirror reflection except the tessellation in the

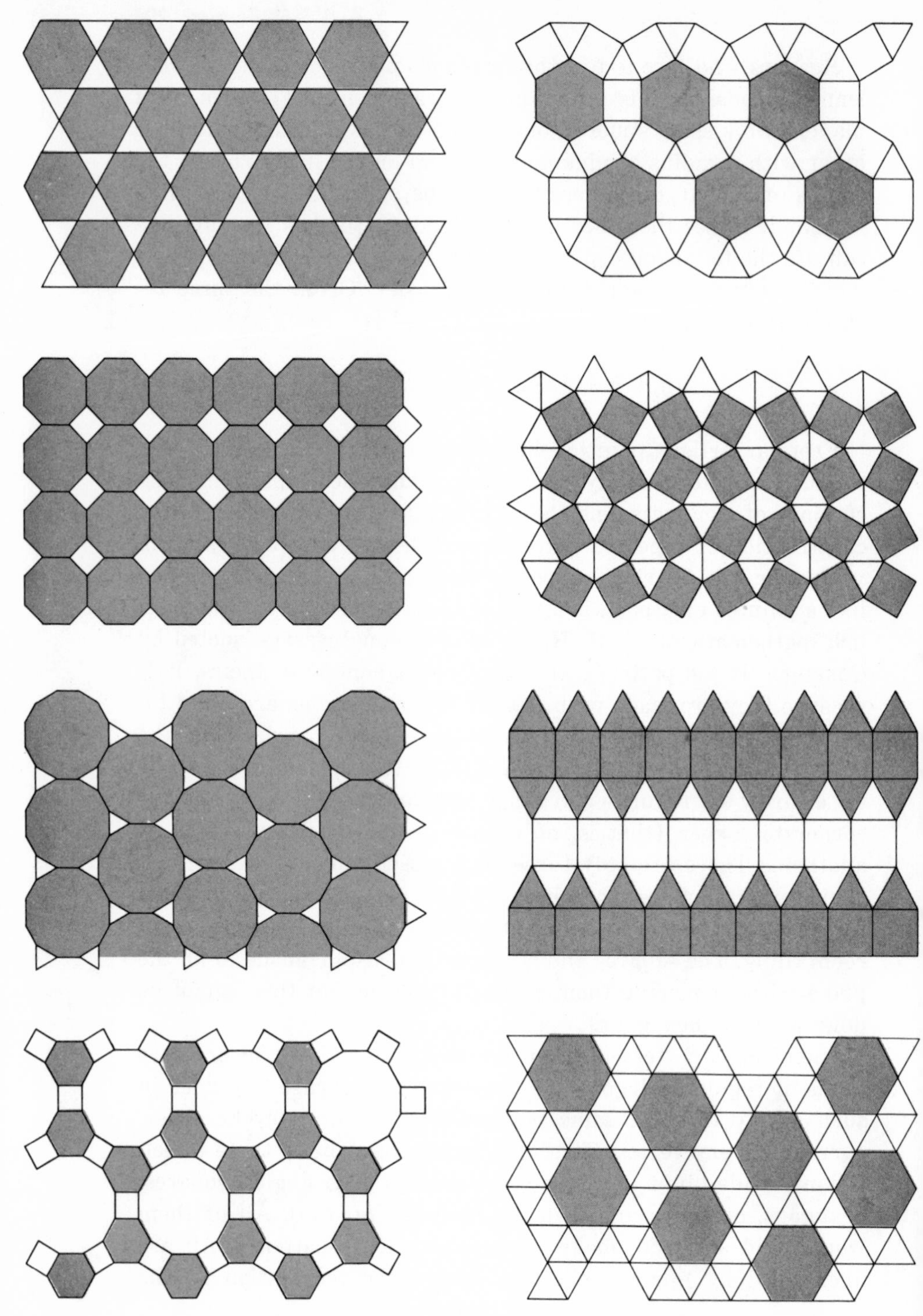

FIG. 105
The eight "semiregular" tessellations.

lower right-hand corner, a pattern first described by Johannes Kepler. It has two forms, each a mirror image of the other. An enjoyable pastime is to cut a large number of cardboard polygons of the required sizes and shapes, paint them various colors and fit them into these tessellations. If the restriction about the vertices is removed, the same polygons will form an infinite variety of mosaics. (Some striking examples of these nonregular but symmetrical tessellations are reproduced in Hugo Steinhaus' *Mathematical Snapshots,* recently reprinted by the Oxford University Press.)

All tessellations that cover the plane with a repeated pattern belong to a set of seventeen different symmetry groups that exhaust all the fundamentally different ways in which patterns can be repeated endlessly in two dimensions. The elements of these groups are simply operations performed on one basic pattern: sliding it along the plane, rotating it or giving it a mirror reversal. The seventeen symmetry groups are of great importance in the study of crystal structure; in fact, Coxeter states that it was the Russian crystallographer E. S. Fedorov who in 1891 first proved that the number of such groups is seventeen. "The art of filling a plane with a repeated pattern," writes Coxeter, "reached its highest development in thirteenth-century Spain, where the Moors used all seventeen groups in their intricate decorations of the Alhambra. Their preference for abstract patterns was due to their strict observance of the Second Commandment ['Thou shalt not make thee any graven image . . .']."

It is not necessary, of course, to limit the fundamental shapes of such patterns to abstract forms. Coxeter goes on to discuss the ingenious way in which the Dutch artist Maurits C. Escher, now living in Baarn, has applied many of the seventeen symmetry groups to mosaics in which animal shapes are used for the fundamental regions. One of Escher's amazing mosaics, reproduced in Coxeter's book, is the knight on horseback shown in Figure 106; another is reproduced in Figure 107. At first glance, Coxeter points out, the knight pattern appears to be the result of sliding a basic shape along horizontal and vertical axes; but on closer inspection one sees that the same basic shape also furnishes the background. Actually, the more interesting symmetry group for this pattern is generated by what are called glide reflections: slid-

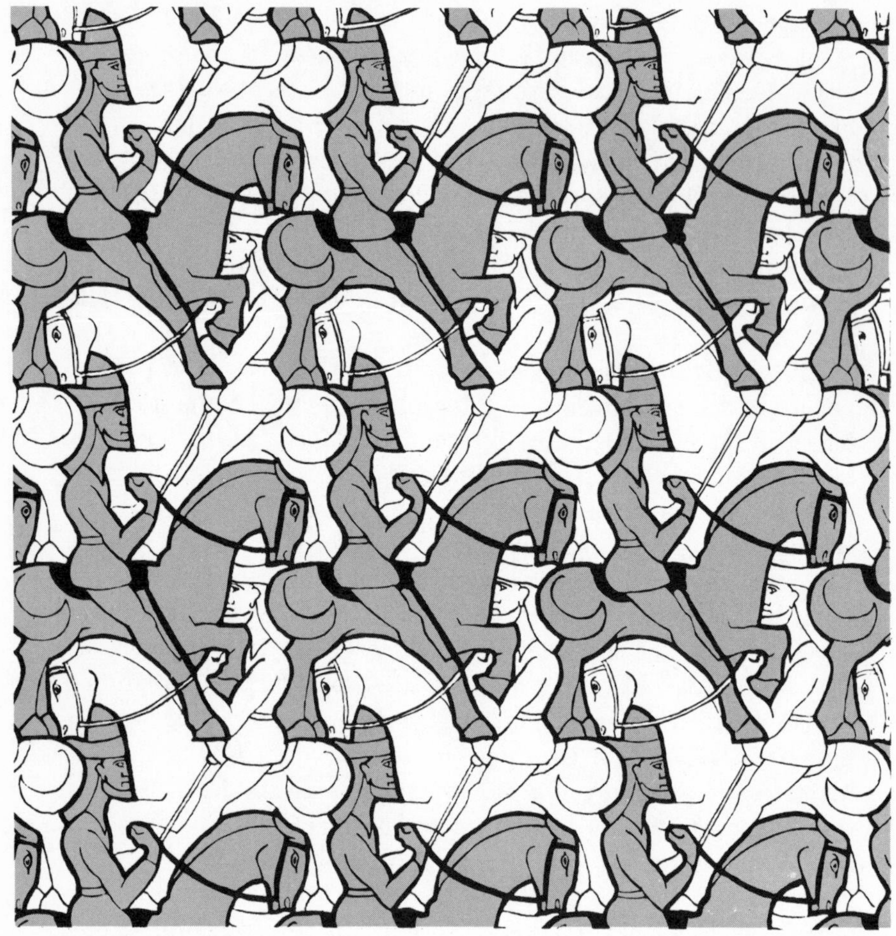

FIG. 106
One of Maurits Escher's mathematical mosaics.

ing the shape and simultaneously giving it a mirror reversal. Strictly speaking, this is not a tessellation because the fundamental region is not a polygon. The pattern belongs to a curious class of mosaics in which irregular shapes, all exactly alike, lock together like pieces in a jigsaw puzzle to cover the plane. Abstract shapes of this sort are not hard to devise, but when they are made to resemble natural objects, they are not so easy to come by.

Escher is a painter who enjoys playing with mathematical structure. There is a respectable school of aesthetics that views all art as a form of play, and an equally respectable school of mathematics that looks upon all mathematical systems as meaningless games played with symbols according to agreed-upon rules. Is science itself another kind of game? On this question Coxeter quotes the following lines from John Lighton Synge, the Irish mathematical physicist:

"Can it be that all the great scientists of the past were really playing a game, a game in which the rules are written not by man

FIG. 107
Another Escher mosaic. It appeared in color on the cover of Scientific American, April 1961.

but by God? . . . When we play, we do not ask why we are playing — we just play. Play serves no moral code except that strange code which, for some unknown reason, imposes itself on the play. . . . You will search in vain through scientific literature for hints of motivation. And as for the strange moral code observed by scientists, what could be stranger than an abstract regard for truth in a world which is full of concealment, deception, and taboos? . . . In submitting to your consideration the idea that the human mind is at its best when playing, I am myself playing, and that makes me feel that what I am saying may have in it an element of truth."

This passage strikes a chord that is characteristic of Coxeter's writings. It is one reason why his book is such a treasure trove for students of mathematics whose minds vibrate on similar wavelengths.

ADDENDUM

GOODRICH COMPANY was not the first to patent a device based on the Moebius strip. Lee De Forest, on January 16, 1923, received patent 1,442,632 for an endless Moebius filmstrip on which sound could be recorded on both sides, and on August 23, 1949, Owen D. Harris received patent 2,479,929 for an abrasive belt in the form of a Moebius band. Readers informed me of both patents; there may be others.

There is an extensive literature on Morley's triangle. Coxeter's proof appears on page 23 of his book, which may be consulted for some earlier references. A full discussion of the triangle, with various other equilateral triangles that turn up (*e.g.*, by trisecting *exterior* angles), is given by W. J. Dobbs in *Mathematical Gazette,* February 1938. The theorem is discussed in H. F. Baker, *Introduction to Plane Geometry,* 1943, pages 345–349. Since Coxeter's book appeared, simple proofs of the theorem have been published by Leon Bankoff, *Mathematics Magazine,* September-October 1962, pages 223-224, and Haim Rose, *American Mathematical Monthly,* August-September 1964, pages 771–773.

The internal bisector problem, known also as the Steiner-Lehmus theorem, has a literature even more vast than the Morley triangle. The theorem was first suggested in 1840 by C. L. Lehmus and first proved by Jacob Steiner. For the problem's fascinating history, and its many solutions, see J. A. McBride, *Edinburgh*

Mathematical Notes, Vol. 33, pages 1–13, 1943, and Archibald Henderson, "The Lehmus-Steiner-Terquem Problem in Global Survey," in *Scripta Mathematica,* Vol. 21, pages 223–312, 1955, and Vol. 22, pages 81–84, 1956. A number of college geometry textbooks prove the theorem: L. S. Shively, *An Introduction to Modern Geometry,* page 141; David R. Davis, *Modern College Geometry,* page 61; Nathan Altshiller-Court, *College Geometry,* page 65. An extremely short proof, by G. Gilbert and D. MacDonnell, appeared in *American Mathematical Monthly,* Vol. 70, page 79, 1963.

Soddy's poem, "The Kiss Precise," is reprinted in its entirety in Clifton Fadiman's entertaining anthology, *The Mathematical Magpie,* Simon and Schuster, 1962, page 284. The last stanza generalizes the theorem to spheres. A fourth stanza, generalizing to hyperspheres of n dimensions, was written by Thorold Gosset and printed in *Nature,* January 9, 1937. This also will be found in Fadiman's book, page 285.

The fourth semiregular tessellation in Fig. 105 (counting left to right) is the basis of a Salvador Dali painting which he calls "Fifty abstract pictures which as seen from two yards change into three Lenines masquerading as Chinese and as seen from six yards appear as the head of a royal tiger." A black-and-white photograph of the painting appeared in *Time,* December 6, 1963, page 90.

Figure 108 reproduces another of Escher's remarkable mosaics: a 1942 lithograph entitled "Verbum." Escher has described it as a pictorial story of creation. "Out of the nebulous grey of the 'Verbum' center ('in the beginning was the Word') triangular figures emerge. The farther they are removed from the center, the sharper becomes the contrast between light and dark, while their original straight outlines become serrated and curved. Alternately, the white becomes background for the black objects and the black for the white objects. Near the edge the figures have evolved into birds, fish and frogs, each species in its proper element: sky, water and earth. At the same time there are gradual transformations from bird into fish, from fish into frog and from frog again into bird. There is a perceptible movement in a clockwise direction." (The quotation is from *The Graphic Work of M. C. Escher,* published in London by the Oldbourne Press, 1961.)

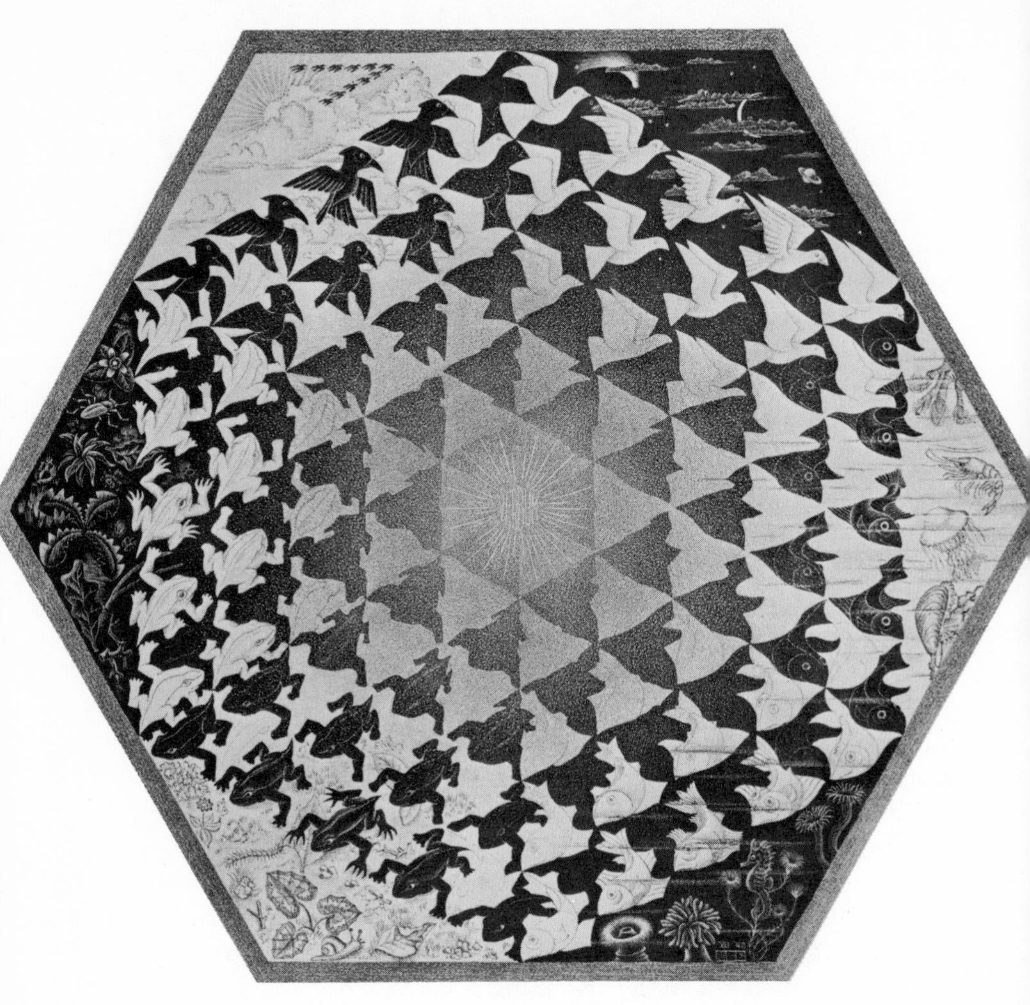

FIG. 108
Escher's "Verbum" (lithograph, 1942). From the collection of Cornelius Van S. Roosevelt, Washington, D.C.

Melvin Calvin, in his article on "Chemical Evolution" in *Interstellar Communication,* edited by A. G. W. Cameron (Benjamin, 1963), reproduces this lithograph, which he says he first saw hanging on the wall of a chemist's office in Holland. "The gradual merging of the figures, one to another," Calvin comments, "and the transformations which eventually become apparent, seem to me to represent the essence not only of life but of the whole universe."

For more on Escher's mathematical art, see my *Scientific American* column for April 1966 and the references it cites.

ANSWERS

READERS were asked to find the radius of the largest sphere that can be placed on a straight line (drawn on a plane) so that it is tangent to two touching spheres, also on the line, with radii of four and nine inches. This can be viewed in cross section [*see Fig. 109*] as a problem involving four mutually tangent circles, the straight line considered a circle of zero curvature. Frederick Soddy's formula for "The Kiss Precise" gives the two circles (drawn with dotted lines) radii of 1 and 11/25 inches and 36 inches respectively. The larger circle is the mid-section of the sphere that answers the problem.

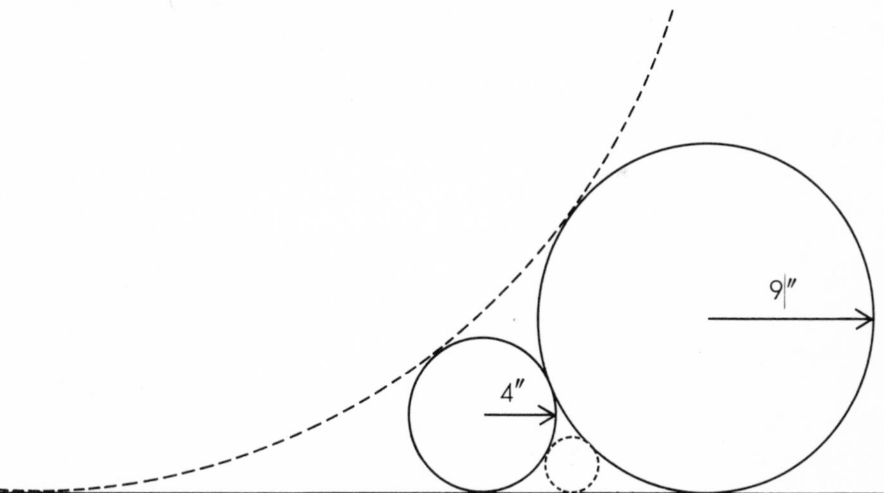

FIG. 109
Answer to the kissing-spheres problem.

CHAPTER EIGHTEEN

□

Bridg-it and Other Games

Man has never shown more ingenuity than in his games.
— Leibniz, in a letter to Pascal

MATHEMATICAL games such as ticktacktoe, checkers, chess and go are contests between two players that (1) must end after a finite number of moves, (2) have no random elements introduced by devices such as dice and cards, (3) are played in such a way that both players see all the moves. If a game is of this type and each player plays "rationally" — that is, according to his best strategy — then the outcome is predetermined. It will be either a draw or a certain win for the player who makes the first move or the player who makes the second move. In this chapter we shall first consider two simple games for which winning strategies are known, then a popular board game for which a winning strategy has just been discovered and a class of board games not yet analyzed.

Many simple games in which pieces are placed on or removed from a board lend themselves to what is called a symmetry strategy. A classic example is the game in which two players take turns placing a domino anywhere on a rectangular board. Each domino must be put down flat, within the border of the rectangle and without moving a previously placed piece. There are enough dominoes to cover the board completely when the pieces are packed side by side. The player who puts down the last domino wins. The game cannot end in a draw, so if both sides play rationally, who is sure to win? The answer is the player who puts down the first domino. His strategy is to place the first domino exactly at the center of the board [*see Fig. 110*] and thereafter to match his opponent's plays by playing symmetrically opposite as shown. It is obvious that whenever the second player finds an open spot, there will always be an open spot to pair with it.

The same strategy applies to any type of flat piece that retains the same shape when it is given a rotation of 180 degrees. For example, the strategy will work if the pieces are Greek crosses; it will not work if they have the shape, say, of the letter T. Will it work if cigars are used as pieces? Yes, but because of the difference in shape between the ends the first cigar must be balanced upright on its flat end! It is easy to invent new games of this sort, in which pieces of different shapes are alternately placed on variously patterned boards according to prescribed rules. In some cases a symmetry strategy provides a win for the first or second player; in other cases no such strategies are possible.

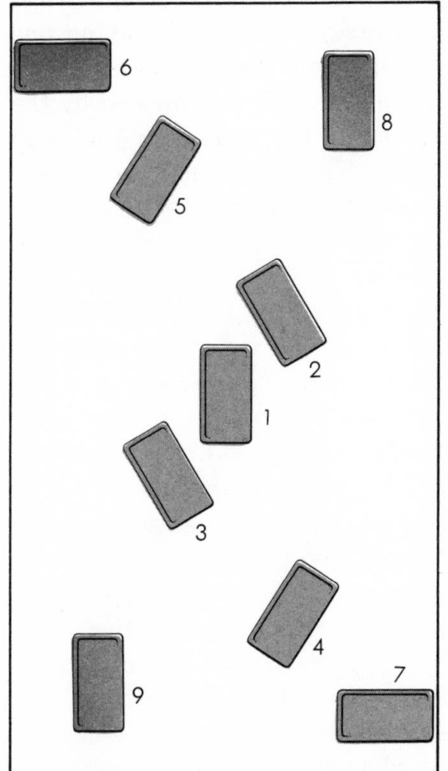

FIG. 110
A domino board game.

A different type of symmetry play wins the following game. Any number of coins are arranged in a circle on the table, each coin touching two of its neighbors. Players alternately remove either one coin or two touching coins. The player who takes the last coin wins. In this case it is the player who makes the second move who can always win. After the player who makes the first move has taken away one or two coins, the remaining coins form a curved chain with two ends. If this chain contains an odd number of coins, the player who makes the second move takes the center coin. If it contains an even number, he takes the two center coins. In both cases he leaves two separate chains of equal length. From this point on, whatever his opponent takes from one chain, he duplicates the move by taking one or two coins from the other chain.

Both this and the preceding strategy are examples of what game theorists sometimes call a pairing strategy: a strategy in which the plays are arranged (not necessarily in symmetrical fashion) in pairs. The optimal strategy consists of playing one member of the pair whenever the opponent plays the other member. A striking example of a pairing strategy is provided by the topological game of Bridg-it, placed on the market in 1960 and now a popular game with children. The reader may remember that Bridg-it was introduced in *Scientific American* in October 1958 as "the game of Gale"; it was devised by David Gale, a mathematician at Brown University.

A Bridg-it board is shown in Figure 111. If it is played on paper, one player uses a black pencil for drawing a straight line to connect any pair of adjacent black spots, horizontally or vertically but not diagonally. The other player uses a red pencil for similarly joining pairs of red spots. Players take turns drawing lines. No line can cross another. The winner is the first player to form a connected path joining the two opposite sides of the board that are his color. (The commercial Bridg-it board has raised spots and small colored plastic bridges that are placed between spots.) For many years a proof has been known that there is a winning strategy for the player who makes the first move, but not until early this year was an actual strategy discovered.

It was Oliver Gross, a games expert in the mathematics department of the Rand Corporation, who cracked the game. When I

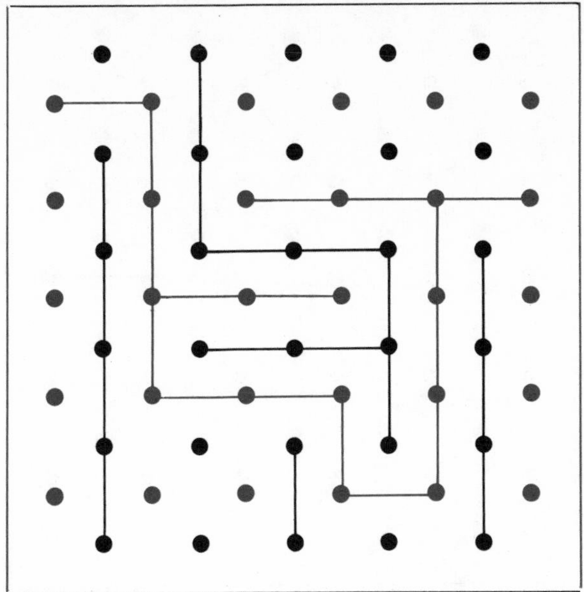

FIG. 111
A finished game of Bridg-it. Red has won.

learned of his discovery, I wrote immediately for details, expecting to receive a long, involved analysis that might prove too technical for this department. To my astonishment the explanation consisted of nothing more than the diagram reproduced in Figure 112 and the following two sentences: Make the first play as indicated by the gray line at lower left in the diagram. Thereafter whenever your opponent's play crosses the end of a dotted line, play by crossing the other end of the same line. This ingenious pairing strategy guarantees a win for the first player, though not necessarily in the fewest moves. Gross describes his strategy as "democratic" in the sense that "it plays stupidly against a stupid opponent, shrewdly against a shrewd one, but wins regardless." This is not the only pairing strategy that Gross discovered, but he picked this one because of its regularity and the ease with which it can be extended to a Bridg-it board of any size.

Note that in the diagram no plays are indicated along the edges of the board. Such plays are allowed by the rules of Bridg-it (in fact, plays of this type are shown on the cover of the box), but there is no point in making such a move, because it can contribute nothing to winning the game. If in the course of playing the winning strategy your opponent throws away a play by making an

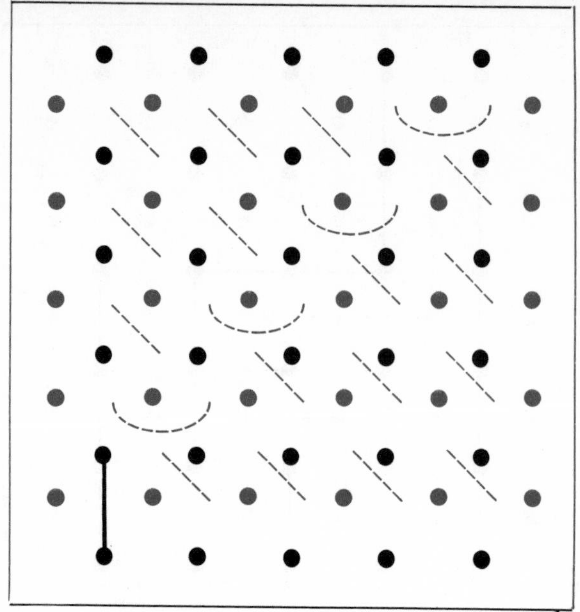

FIG. 112
Oliver Gross's pairing strategy for winning at Bridg-it.

edge move, you can counter with an edge move of your own. Or, if you prefer, you can play *anywhere* on the board. If at some point later in the game this random move is demanded by the strategy, you simply play somewhere else. Having an extra play on the board is sometimes an asset, never a liability. Of course, now that a winning strategy for the first player is known, Bridg-it ceases to be of interest except to players who have not yet heard the news.

Many board games with relatively simple rules have defied all attempts at mathematical analysis. An example is provided by the family of games that derives from halma, a game widely played in England late in the 19th century. "The normal English way," wrote George Bernard Shaw in 1898, is "to sit in separate families in separate rooms in separate houses, each person silently occupied with a book, a paper, or a game of halma. . . ." (This quotation is given in *The New Complete Hoyle,* by Albert H. Morehead, Richard L. Frey and Geoffrey Mott-Smith.)

The original halma (the name is a Greek word for "leap") was played on a checkerboard with sixteen squares to a side, but the basic mode of play was soon extended to other boards of varying size and shape. The game known today as Chinese checkers is one of the many later varieties of halma. I shall explain here only a simplified version, which can be played on the familiar eight-by-eight checkerboard and which leads to an entertaining solitaire puzzle that is still unsolved.

The game begins with the checkers in the standard starting position for a checker game. Moves are the same as in checkers, with these exceptions:

1. No jumped pieces are removed.
2. A checker may jump men of either color.
3. Backward moves and jumps are permitted.

A chain of unbroken jumps may be made over men of both colors, but one is not allowed to combine jumps with a nonjump move. The object of the game for each player is to occupy his opponent's starting position. The first to do so is the winner. A player also wins if the game reaches a situation in which his opponent is unable to move.

Some notion of how difficult it is to analyze games of the halma type can be had by working on the following puzzle. Arrange twelve checkers in the usual starting positions on the black squares of the first three rows of a checkerboard. The rest of the board is empty. In how few halma plays can you transport these men to the three rows on the opposite side of the board? A "play" is defined as either a diagonal checker move, forward or back, to a neighboring black square; or a jump over one or more men. An unbroken jump may include forward and backward leaps and is counted as a single play. As in halma, it is not compulsory to jump when jumps are available, and a series of unbroken jumps may be terminated wherever desired, even though more jumps are possible.

For convenience in recording a solution, number the black squares, left to right and top to bottom, from 1 to 32.

ADDENDUM

AFTER THE twenty-move solution of the checker problem was published, several readers sent proofs that at least eighteen moves were required. One reader, Vern Poythress, Fresno, California, sent a twenty-move-minimum proof; unfortunately, too long and involved to give here.

As I pointed out in *The 2nd Scientific American Book of Mathematical Puzzles & Diversions,* Bridg-it is identical with a switching game called Bird Cage that was invented by Claude E. Shannon. The Shannon game is described in one of Arthur Clarke's short stories, "The Pacifist," reprinted in Clifton Fadiman's anthology, *Mathematical Magpie* (Simon and Schuster, 1962), pages 37–47; and in Marvin Minsky, "Steps Toward Artificial Intelligence," *Proceedings of the Institute of Radio Engineers*, Vol. 49, 1961, page 23. In addition to Bridg-it, manufactured by Hassenfeld Brothers, there is now a more expensive version of the game on the market under the name of Twixt, put out by 3M Brand Bookshelf Games.

Independently of Gross's work, a winning strategy for Bridg-it was discovered by Alfred Lehman, of the U.S. Army's Mathematical Research Center, University of Wisconsin. Lehman found a general strategy for a wide class of Shannon switching games, of which Bird Cage (or Bridg-it) is one species. Lehman wrote me that he first worked out his system in March 1959, and although it was mentioned in a Signal Corps report and in an outline sent to Shannon, it was not then published. In April 1961 he spoke about it at a meeting of the American Mathematical Association, a summary of his paper appearing in the association's June notices. A full, formal presentation, "A Solution of the Shannon Switching Game," was published in the *Journal of the Society of Industrial Applied Mathematics*, Vol. 12, No. 4, December 1964, pages 687–725. Lehman's strategy comes close to providing a winning strategy for Hex, a well-known topology game similar to Bridg-it, but Hex slipped through the analysis and remains unsolved.

In 1961 Günter Wenzel wrote a Bridg-it-playing program for the IBM 1401 computer, basing it on the Gross strategy. His description of the program was issued as a photocopied typescript by the IBM Systems Research Institute, New York City, and in 1963 was published in Germany in the March issue of *Bürotechnik und Automation*.

ANSWERS

THE PROBLEM of moving twelve checkers from one side of the board to the other, using halma moves, brought a heavy response from readers. More than 30 readers solved the problem in 23 moves, 49 solved it in 22 moves, 31 in 21 moves and 14 in 20 moves. The fourteen winners, in the order their letters are dated, are: Edward J. Sheldon, Lexington, Massachusetts; Henry Laufer, New York City; Donald Vanderpool, Towanda, Pennsylvania; Corrado Böhm and Wolf Gross, Rome, Italy; Otis Shuart, Syracuse, New York; Thomas Storer, Melrose, Florida; Forrest Vorks, Seattle, Washington; Georgianna March, Madison, Wisconsin; James Burrows, Stanford, California; G. W. Logemann, New York City; John Stout, New York City; Robert Schmidt, State College, Pennsylvania; G. L. Lupfer, Solon, Ohio; and J. R. Bird, Toronto, Canada.

No proof that twenty is the minimum was received, although many readers indicated a simple way to prove that at least sixteen moves are required. At the start, eight checkers are on odd rows 1 and 3, four checkers on even row 2. At the finish, eight checkers are on even rows 6 and 8, four checkers on odd row 7. Clearly four checkers must change their parity from odd to even. This can be done only if each of the four makes at least one jump move and one slide move, thereby bringing the total of required moves to sixteen.

It is hard to conceive that the checkers could be transported in fewer than twenty moves, although I must confess that when I presented the problem I found it equally hard to conceive that it could be solved in as few as twenty moves. Assuming that the black squares are numbered 1 to 32, left to right and top to bottom, with a red square in the board's upper left corner, Sheldon's twenty-move solution (the first answer to be received) is as follows:

1.	21–17	11.	14–5
2.	30–14	12.	23–7
3.	25–9	13.	18–2
4.	29–25	14.	32–16
5.	25–18	15.	27–11
6.	22–6	16.	15–8
7.	17–1	17.	8–4
8.	31–15	18.	24–8
9.	26–10	19.	19–3
10.	28–19	20.	16–12

This solution is symmetrical. Figure 113 shows the position of the checkers after the tenth move. If the board is now inverted and the first ten moves are repeated in reverse order, the transfer is completed. So far as I know, this is the first published solution in twenty moves. It is far from unique. Other symmetrical twenty-move solutions were received, along with one wildly asymmetrical one from Mrs. March, the only woman reader to achieve the minimum.

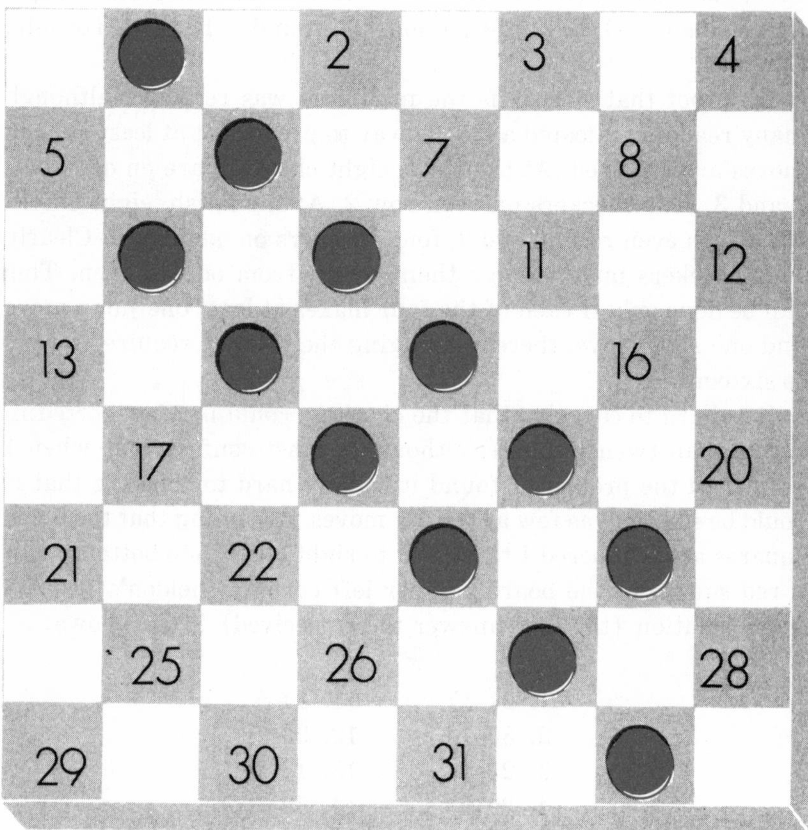

FIG. 113
Position of checkers after ten moves.

CHAPTER NINETEEN

□

Nine More Problems

1. COLLATING THE COINS

ARRANGE THREE pennies and two dimes in a row, alternating the coins as shown in Figure 114. The problem is to change their positions to those shown at the bottom of the illustration in the shortest possible number of moves.

A move consists of placing the tips of the first and second fingers on any two touching coins, *one of which must be a penny and the other a dime,* then sliding the pair to another spot along the imaginary line shown in the illustration. The two coins in the pair must touch at all times. The coin at left in the pair must remain at left; the coin at right must remain at right. Gaps in the chain are allowed at the end of any move except the final one. After the last move the coins need not be at the same spot on the imaginary line that they occupied at the start.

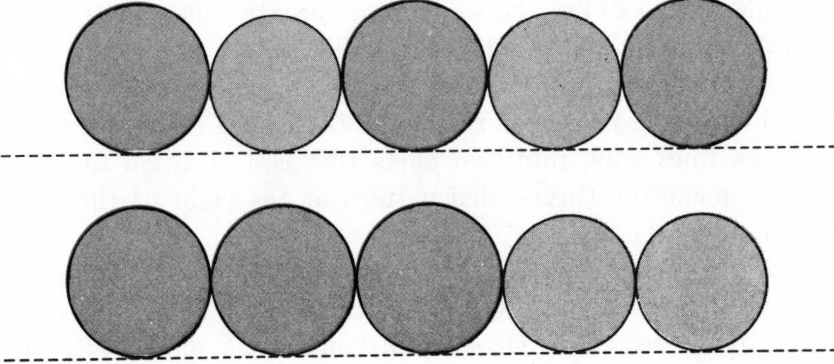

FIG. 114
The pennies and dimes puzzle.

If it were permissible to shift two coins of the same kind, the puzzle could be solved easily in three moves: slide 1, 2 to left, fill the gap with 4, 5, then move 5, 3 from right to left end. But with the proviso that each shifted pair must include a dime and penny it is a baffling and pretty problem. H. S. Percival, of Garden City, New York, was the first to call it to my attention.

2. TIME THE TOAST

EVEN THE SIMPLEST of household tasks can present complicated problems in operational research. Consider the preparation of three seconds to take it out and three seconds to reverse a slice type, with hinged doors on its two sides. It holds two pieces of bread at once but toasts each of them on one side only. To toast both sides it is necessary to open the doors and reverse the slices.

It takes three seconds to put a slice of bread into the toaster, three seconds to take it out and three seconds to reverse a slice without removing it. Both hands are required for each of these operations, which means that it is not possible to put in, take out or turn two slices simultaneously. Nor is it possible to butter a slice while another slice is being put into the toaster, turned, or taken out. The toasting time for one side of a piece of bread is thirty seconds. It takes twelve seconds to butter a slice.

Each slice is buttered on one side only. No side may be buttered until it has been toasted. A slice toasted and buttered on one side may be returned to the toaster for toasting on its other side. The toaster is warmed up at the start. In how short a time can three slices of bread be toasted on both sides and buttered?

3. TWO PENTOMINO POSERS

FOR PENTOMINO BUFFS, here are two recently discovered problems, the first one easy and the second difficult.

A. At the left of Figure 115 the twelve pentominoes are arranged to form a six-by-ten rectangle. Divide the rectangle, along the black lines only, into two parts that can be fitted together again to make the three-holed pattern at the right of the illustration.

B. Arrange the twelve pentominoes to form a six-by-ten rectangle but in such a way that each pentomino touches the border of the rectangle. Of several thousand fundamentally different ways of making the six-by-ten rectangle (rotations and

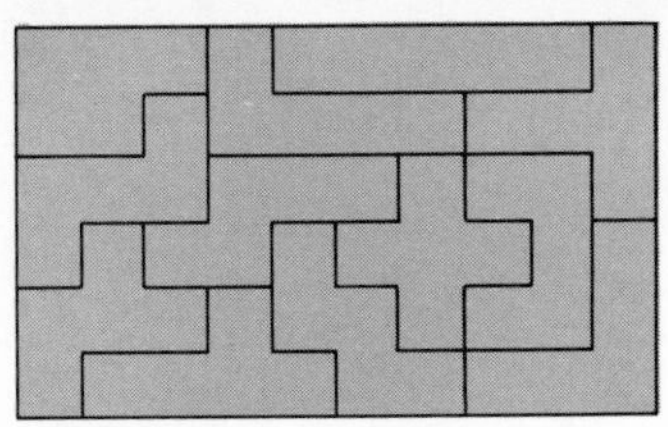

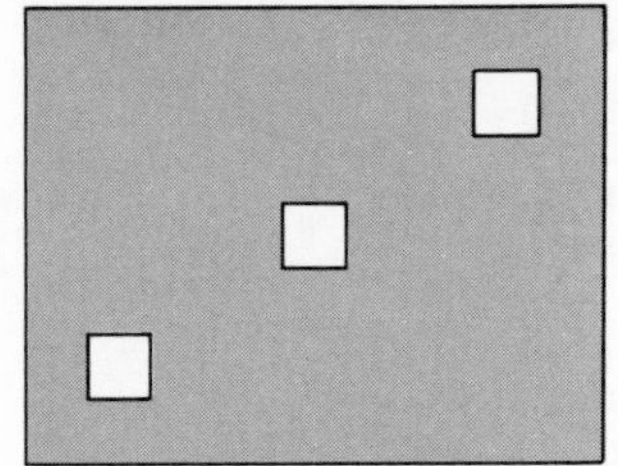

FIG. 115
A pentomino problem.

reflections are not considered different), only two are known to meet the condition of this problem. Asymmetrical pieces may be turned over and placed with either side against the table.

4. A FIXED-POINT THEOREM

ONE MORNING, exactly at sunrise, a Buddhist monk began to climb a tall mountain. The narrow path, no more than a foot or two wide, spiraled around the mountain to a glittering temple at the summit.

The monk ascended the path at varying rates of speed, stopping many times along the way to rest and to eat the dried fruit he carried with him. He reached the temple shortly before sunset. After several days of fasting and meditation he began his journey back along the same path, starting at sunrise and again walking at variable speeds with many pauses along the way. His average speed descending was, of course, greater than his average climbing speed.

Prove that there is a spot along the path that the monk will occupy on both trips at precisely the same time of day.

5. A PAIR OF DIGIT PUZZLES

THE FOLLOWING two problems seem to call for a digital computer so that hundreds of combinations of digits can be tested in a reasonable length of time. But if approached properly and with the aid of a clever dodge or two, both problems can be solved with very little pencil and paper work. It is by such short cuts that a skillful programmer often can save his company valuable computer time and in some cases even eliminate a need for the computer.

A. *The Square Root of Wonderful* was the name of a recent play on Broadway. If each letter in WONDERFUL stands for a different digit (zero excluded) and if OODDF, using the same code, represents the square root, then what *is* the square root of wonderful?

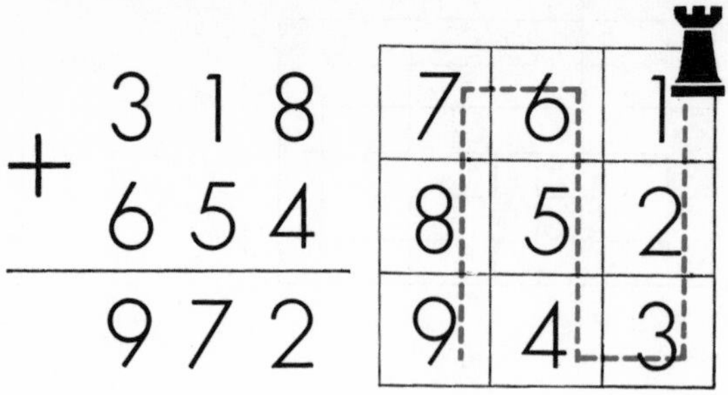

	3	1	8
+	6	5	4
	9	7	2

7	6	1
8	5	2
9	4	3

FIG. 116
Can features of both squares be combined?

B. There are many ways in which the nine digits (not counting zero) can be arranged in square formation to represent a sum. In the example shown at left in Figure 116, 318 plus 654 equals 972. There are also many ways to place the digits on a square matrix so that, taken in serial order, they form a rookwise connected chain. An example is at right in the illustration. You can start at 1, then, moving like a chess rook, one square per move, you can advance to 2, 3, 4 and so on to 9.

The problem is to combine both features in the same square. In other words, place the digits on a three-by-three matrix so that they form a rookwise connected chain, from 1 to 9, and also in such a way that the bottom row is the sum of the first two rows. The answer is unique.

6. HOW DID KANT SET HIS CLOCK?

It is said that Immanuel Kant was a bachelor of such regular habits that the good people of Königsberg would adjust their clocks when they saw him stroll past certain landmarks.

One evening Kant was dismayed to discover that his clock had run down. Evidently his manservant, who had taken the day off, had forgotten to wind it. The great philosopher did not reset the hands because his watch was being repaired and he had no way of knowing the correct time. He walked to the home of his friend Schmidt, a merchant who lived a mile or so away, glancing at the clock in Schmidt's hallway as he entered the house.

After visiting Schmidt for several hours Kant left and walked home along the route by which he came. As always, he walked with a slow, steady gait that had not varied in twenty years. He had no notion of how long this return trip took. (Schmidt had recently moved into the area and Kant had not yet timed himself on this walk.) Nevertheless, when Kant entered his house, he immediately set his clock correctly.

How did Kant know the correct time?

7. PLAYING TWENTY QUESTIONS WHEN PROBABILITY VALUES ARE KNOWN

IN THE well-known game Twenty Questions one person thinks of an object, such as the Liberty Bell in Philadelphia or Lawrence Welk's left little toe, and another person tries to guess the object by asking no more than twenty questions, each answerable by yes or no. The best questions are usually those that divide the set of possible objects into two subsets as nearly equal in number as possible. Thus if a person has chosen as his "object" a number from 1 through 9, it can be guessed by this procedure in no more than four questions — possibly less. In twenty questions one can guess any number from 1 through 2^{20} (or 1,048,576).

Suppose that each of the possible objects can be given a different value to represent the probability that it has been chosen. For example, assume that a deck of cards consists of one ace of spades, two deuces of spades, three threes, and on up to nine nines, making 45 spade cards in all. The deck is shuffled; someone draws a card. You are to guess it by asking yes-no questions. How can you minimize the number of questions that you will probably have to ask?

8. DON'T MATE IN ONE

KARL FABEL, a German chess problemist, is responsible for the outrageous problem depicted in Figure 117. It appeared recently

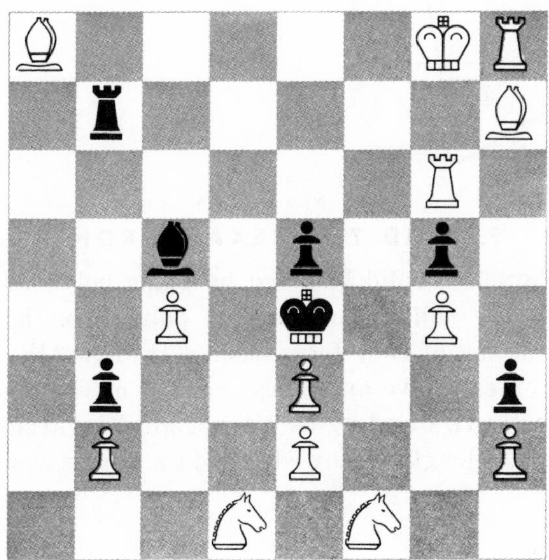

FIG. 117
White to move and not checkmate.

in Mel Stover's delightful column of offbeat chess puzzles in *Canadian Chess Chat* magazine.

You are asked to find a move for white that will *not* result in an immediate checkmate of the black king.

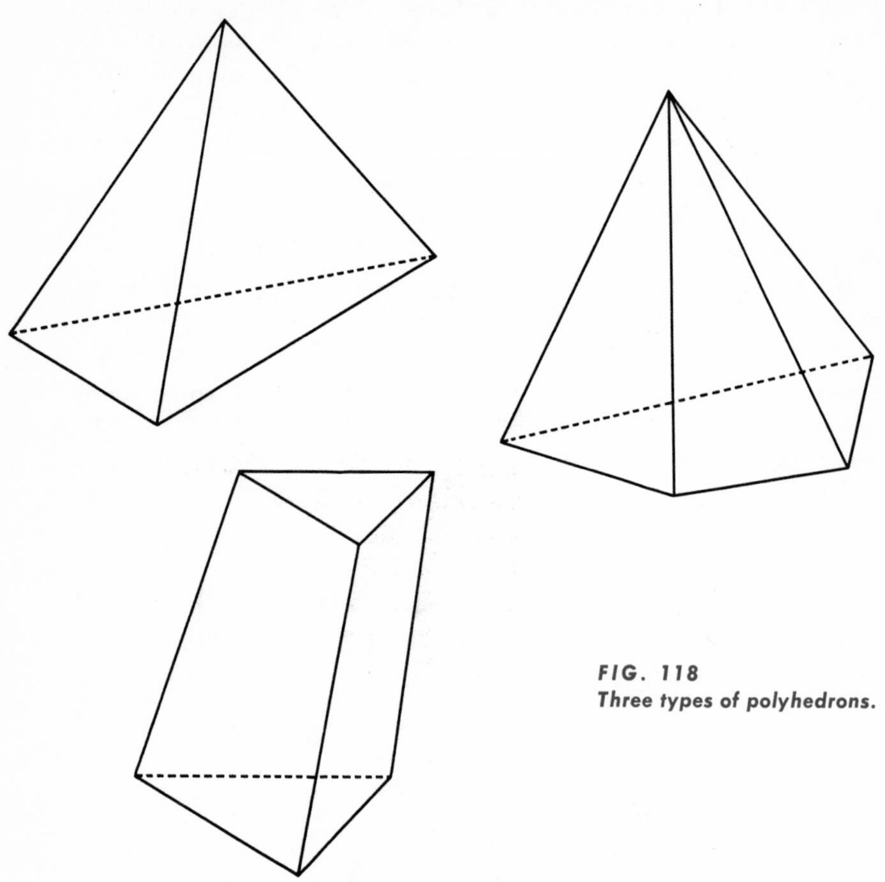

FIG. 118
Three types of polyhedrons.

9. FIND THE HEXAHEDRONS

A POLYHEDRON is a solid bounded by plane polygons known as the faces of the solid. The simplest polyhedron is the tetrahedron, consisting of four faces, each a triangle [*Fig. 118, top*]. A tetrahedron can have an endless variety of shapes, but if we regard its network of edges as a topological invariant (that is, we may alter the length of any edge and the angles at which edges meet but we must preserve the structure of the network), then there is only one basic type of tetrahedron. It is not possible, in other words, for a tetrahedron to have sides that are anything but triangles.

The five-sided polyhedron has two basic varieties [*Fig. 118, middle and bottom*]. One is represented by the Great Pyramid of Egypt (four triangles on a quadrilateral base). The other is represented by a tetrahedron with one corner sliced off; three of its faces are quadrilaterals, two are triangles.

John McClellan, an artist in Woodstock, New York, asks this question: How many basic varieties of convex hexahedrons, or six-sided solids, are there altogether? (A solid is convex if each of its sides can be placed flat against a table top.) The cube is, of course, the most familiar example.

If you search for hexahedrons by chopping corners from simpler solids, you must be careful to avoid duplication. For example, the Great Pyramid, with its apex sliced off, has a skeleton that is topologically equivalent to that of the cube. Be careful also to avoid models that cannot exist without warped faces.

ANSWERS

1. The dime and penny puzzle can be solved in four moves as follows. Coins are numbered from left to right.

1. Move 3, 4 to the right of 5 but separated from 5 by a gap equal to the width of two coins.
2. Move 1, 2 to the right of 3, 4, with coins 4 and 1 touching.
3. Move 4, 1 to the gap between 5 and 3.
4. Move 5, 4 to the gap between 3 and 2.

2. Three slices of bread — A, B, C — can be toasted and buttered on the old-fashioned toaster in two minutes. Figure 119 shows the way to do it.

After this solution appeared, I was staggered to hear from five readers that the time could be cut to 111 seconds. What I had overlooked was the possibility of partially toasting one side of a slice, removing it, then returning it later to complete the toasting. Solutions of this type arrived from Richard A. Brouse, a pro-

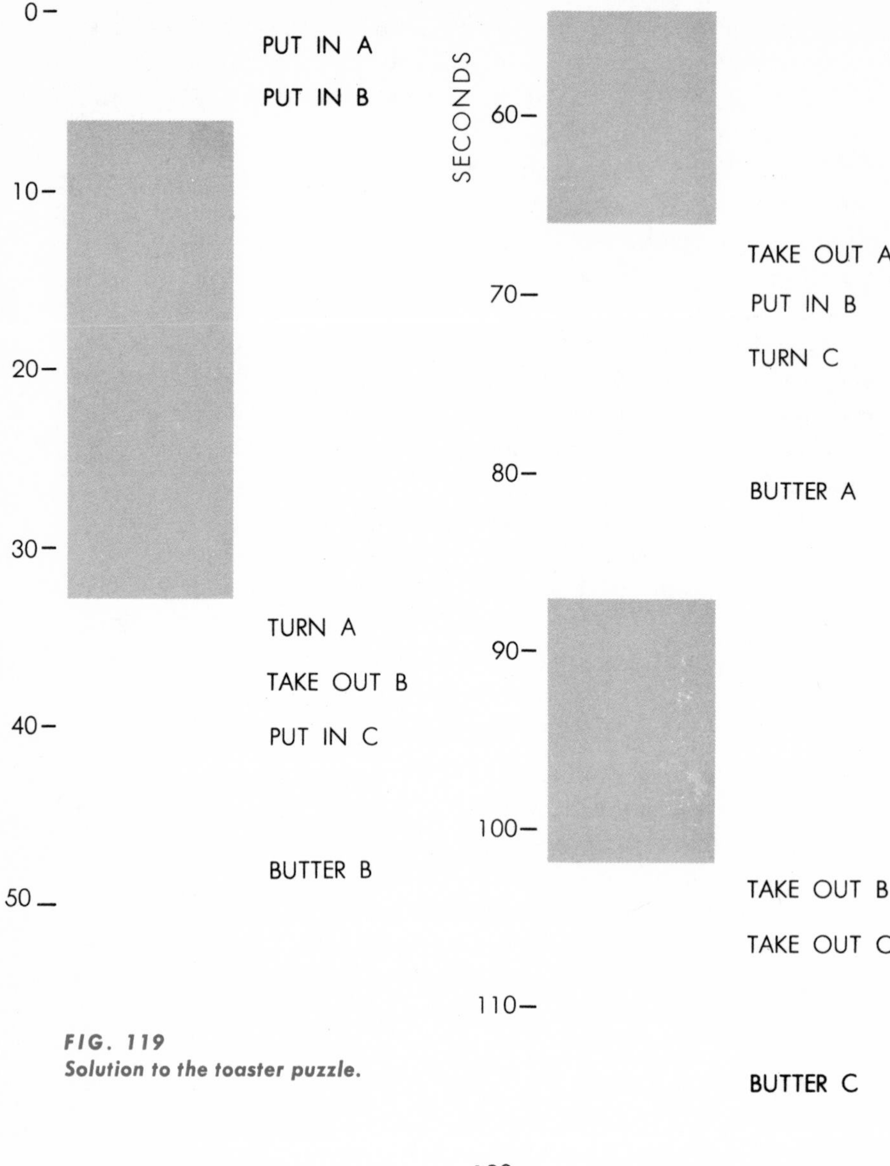

FIG. 119
Solution to the toaster puzzle.

graming systems analyst with IBM, San Jose, California; R. J. Davis, Jr., of General Precision Inc., Little Falls, New Jersey; John F. O'Dowd, Quebec; Mitchell P. Marcus, Binghamton, New York; and Howard Robbins, Vestal, New York.

Davis' procedure is as follows:

Seconds	*Operation*
1–3	Put in slice A.
3–6	Put in B.
6–18	A completes 15 seconds of toasting on one side.
18–21	Remove A.
21–23	Put in C.
23–36	B completes toasting on one side.
36–39	Remove B.
39–42	Put in A, turned.
42–54	Butter B.
54–57	Remove C.
57–60	Put in B.
60–72	Butter C.
72–75	Remove A.
75–78	Put in C.
78–90	Butter A.
90–93	Remove B.
93–96	Put in A, turned to complete the toasting on its partially toasted side.
96–108	A completes its toasting.
108–111	Remove C.

All slices are now toasted and buttered, but slice A is still in the toaster. Even if A must be removed to complete the entire operation, the time is only 114 seconds.

Robbins pointed out that near the end, while A is finishing its toasting, one can use the time efficiently by eating slice B.

3. Figure 120 shows how the six-by-ten rectangle, formed with the twelve pentominoes, can be cut into two parts and the parts

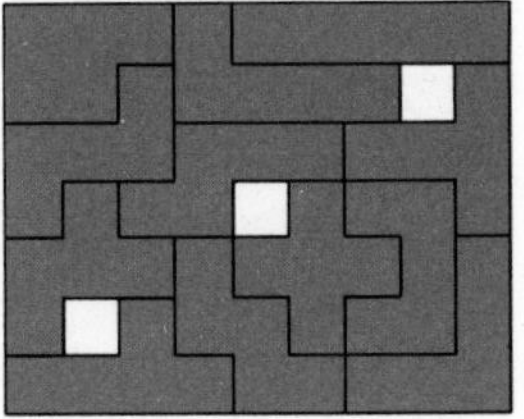

FIG. 120
A six-by-ten rectangle made up of pentominoes is refitted into a seven-by-nine one with three holes.

refitted to make the seven-by-nine rectangle with three interior holes. Figure 121 shows the only two possible patterns for the six-by-ten rectangle in which all twelve pieces touch the border. The first of these patterns is also remarkable in that it can be divided (like the rectangle in the preceding pentomino problem) into two congruent halves.

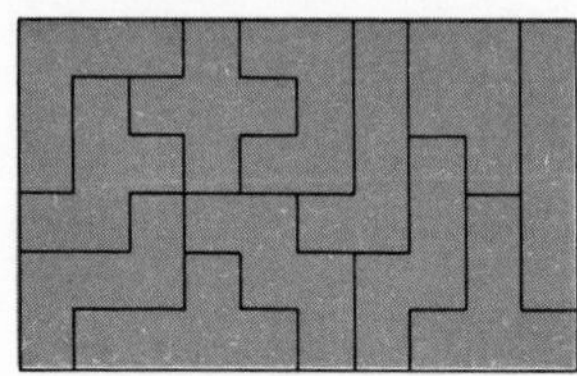

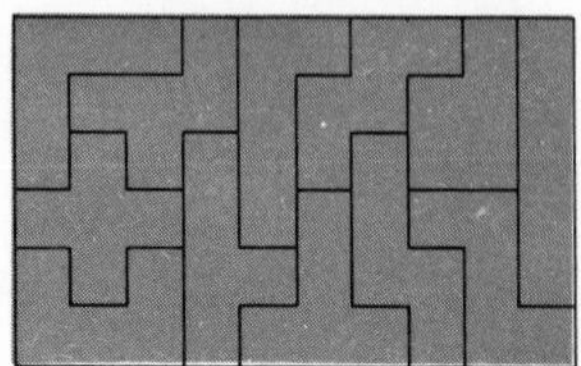

FIG. 121
All the pentominoes in these six-by-ten rectangles touch the border of the rectangle.

4. A man goes up a mountain one day, down it another day. Is there a spot along the path that he occupies at the same time of day on both trips? This problem was called to my attention by psychologist Ray Hyman, of the University of Oregon, who in turn found it in a monograph entitled "On Problem-Solving," by the German Gestalt psychologist Karl Duncker. Duncker writes of being unable to solve it and of observing with satisfaction that others to whom he put the problem had the same difficulty. There are several ways to go about it, he continues, "but probably none is . . . more drastically evident than the following. Let ascent and descent be divided between *two* persons on the same day. They must *meet*. Ergo. . . . With this, from an unclear dim condition not easily surveyable, the situation has suddenly been brought into full daylight."

5. A. If OODDF is the square root of WONDERFUL, what number does it represent? O cannot be greater than 2 because this would give a square of ten digits. It cannot be 1 because there is no way that a number, beginning with 11, can have a square in which the second digit is 1. Therefore O must be 2.

WONDERFUL must be between the squares of 22,000 and 23,000. The square of 22 is 484; the square of 23 is 529. Since the second digit of WONDERFUL is 2, we conclude that WO = 52.

What values for the letters of 22DDF will make the square equal 52NDERFUL? The square of 229 is 52,441; the square of 228 is 51,984. Therefore OODD is either 2,299 or 2,288.

We now use a dodge based on the concept of digital root. The sum of the nine digits in WONDERFUL (we were told zero is excluded) is 45, which in turn sums to 9, its digital root. Its square root must have a digital root that, when squared, gives a number with a digital root of nine. The only digital roots meeting this requirement are 3, 6, 9, therefore OODDF must have a digital root of 3, 6 or 9.

F cannot be 1, 5 or 6, because any of those digits would put an F at the end of WONDERFUL. The only possible completions of 2299F and 2288F that meet the digital root requirement are 22,998, 22,884 and 22,887.

The square of 22,887 is 523,814,769, the only one that fits the code word WONDERFUL.

B. The timesaving insight in this problem is the realization that if the nine digits are placed on a three-by-three matrix to form a rookwise connected chain from 1 to 9, the odd digits must occupy the central and four corner cells. This is easily seen by coloring the nine cells like a checkerboard, the center cell black. Since there is one more black cell than white, the path must begin and end on black cells, and all even digits will fall on white cells.

There are 24 different ways in which the four even digits can be arranged on the white cells. Eight of these, in which 2 is opposite 4, can be eliminated immediately because they do not permit a complete path of digits in serial order. The remaining sixteen patterns can be quickly checked, keeping in mind that the sum of the two upper digits on the left must be less than 10 and the sum of the two upper digits on the right must be more than 10. The second assertion holds because the two upper digits in the middle are even and odd, yet their sum is an even digit. This could happen only if 1 is carried over from the sum of the right column. The only way to form the path so that the bottom row of the square is the sum of the first and second rows is shown in Figure 122.

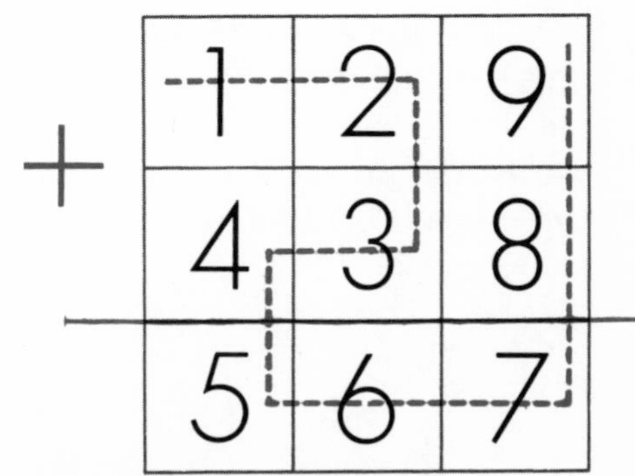

FIG. 122
Solution to the chain-of-digits problem.

When this solution appeared in *Scientific American,* Harmon H. Goldstone, New York City, and Scott B. Kilner, Corona, California, wrote to explain a faster method they had used. There are only three basically different rook paths (ignoring rotations and reflections) on the field: the one shown in the solution, a spiral path from corner to center, and an "S" path from corner to diagonally opposite corner. On each path the digits can run in order in either direction, making six different patterns. By considering each in its various rotations and reflections, one quickly arrives at the unique answer.

Note that if the solution is mirror inverted (by a mirror held above it), it forms a square, its digits still in rookwise serial order, such that the middle row subtracted from the top row gives the bottom row.

Charles W. Trigg, in a detailed analysis of solutions to ABC + DEF = GHK (in *Recreational Mathematics Magazine,* No. 7, February 1962, pages 35–36), gives the only three solutions, in addition to the one shown here, on which the digits 1 through 9 are in serial order along a *queenwise* connected path.

6. Immanuel Kant calculated the exact time of his arrival home as follows. He had wound his clock before leaving, so a glance at its face told him the amount of time that had elapsed during his absence. From this he subtracted the length of time spent with Schmidt (having checked Schmidt's hallway clock when he arrived and again when he left). This gave him the total time spent in walking. Since he returned along the same route, at the same speed, he halved the total walking time to obtain the length of time it took him to walk home. This added to the time of his departure from Schmidt's house gave him the time of his arrival home.

Winston Jones, of Johannesburg, South Africa, wrote to suggest another solution. Mr. Schmidt, in addition to being Kant's friend, was also his watchmaker. So while Kant sat and chatted with him, he repaired Kant's watch.

7. The first step is to list in order the probability values for the nine cards: 1/45, 2/45, 3/45. . . . The two lowest values are combined to form a new element: 1/45 plus 2/45 equals 3/45. In other words, the probability that the chosen card is either an ace or deuce is 3/45. There are now eight elements: the ace-deuce set, the three, the four, and so on up to nine. Again the two lowest probabilities are combined: the ace-deuce value of 3/45 and the 3/45 probability that the card is a three. This new element, consisting of aces, deuces and threes, has a probability value of 6/45. This is greater than the values for either the fours or fives, so when the two lowest values are combined again, we must pair the fours and fives to obtain an element with the value of 9/45. This procedure of pairing the lowest elements is continued until only one element remains. It will have the probability value of 45/45, or 1. The chart in Figure 123 shows how the elements are combined. The strategy for minimizing the number of questions is to take these pairings in reverse order. Thus the first question could be: Is the card in the set of fours, fives and nines? If not, you know it is in the other set so you ask next: Is it a seven or eight? And so on until the card is guessed.

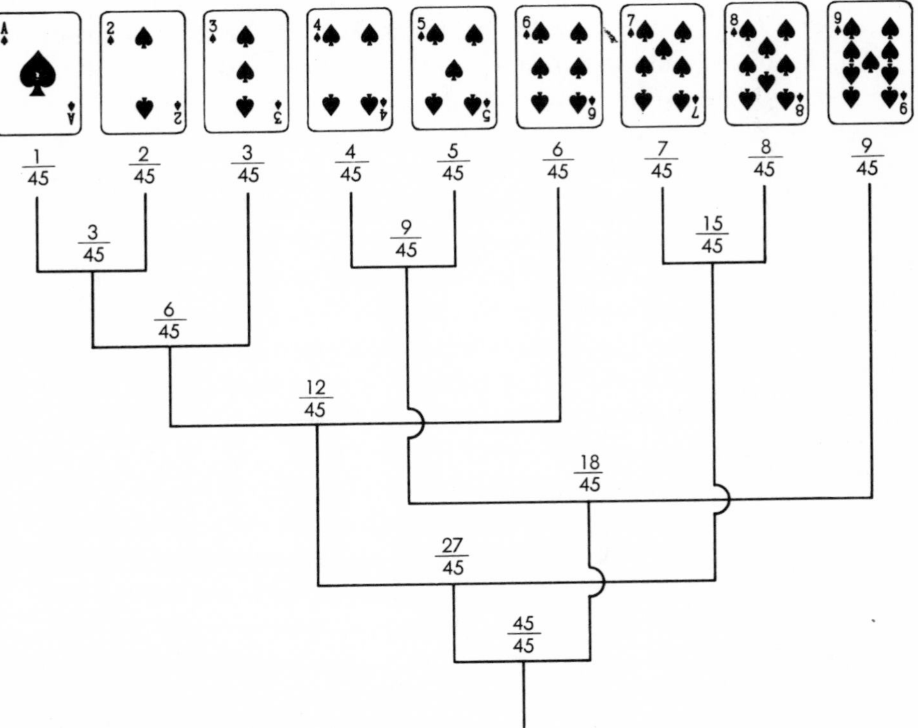

FIG. 123
Strategy for minimizing the number of yes-no questions in guessing one of a number of objects with probability values.

Note that if the card should be an ace or deuce it will take five questions to pinpoint it. A binary strategy, of simply dividing the elements as nearly as possible into halves for each question, will ensure that no more than four questions need be asked, and you might even guess the card in three. Nevertheless, the previously described procedure will give a slightly lower expected minimum number of questions in the long run; in fact, the lowest possible. In this case, the minimum number is three.

The minimum is computed as follows: Five questions are needed if the card is an ace. Five are also needed if the card is a deuce, but there are two deuces, making ten questions in all. Similarly, the three threes call for three times four, or twelve, questions. The total number of questions for all 45 cards is 135, or an average of three questions per card.

This strategy was first discovered by David A. Huffman, an electrical engineer at M.I.T., while he was a graduate student there. It is explained in his paper "A Method for the Construction of Minimum-Redundancy Codes," *Proceedings of the Institute of Radio Engineers,* Vol. 40, pages 1098–1101, September 1952. It was later rediscovered by Seth Zimmerman, who described it in his article on "An Optimal Search Procedure," *American Mathematical Monthly,* Vol. 66, pages 690–693, October 1959. A good nontechnical exposition of the procedure will be found in John R. Pierce, *Symbols, Signals and Noise* (Harper & Brothers, 1961), beginning on page 94.

8. In the chess problem white can avoid checkmating black only by moving his rook four squares to the west. This checks the black king, but black is now free to capture the checking bishop with his rook.

When this problem appeared in *Scientific American,* dozens of readers complained that the position shown is not a possible one because there are two white bishops on the same color squares. They forgot that a pawn on the last row can be exchanged for any piece, not just the queen. Either of the two missing white pawns could have been promoted to a second bishop.

There have been many games by masters in which pawns were promoted to knights. Promotions to bishops are admittedly rare, yet one can imagine situations in which it would be desirable. For instance, to avoid stalemating the opponent. Or white may see that he can use either a new queen or a new bishop in a subtle checkmate. If he calls for a queen, it will be taken by a black rook, in turn captured by a white knight. But if white calls for a bishop, black may be reluctant to trade a rook for bishop and so let the bishop remain.

9. The seven varieties of convex hexahedrons, with topologically distinct skeletons, are shown in Fig. 124. I know of no simple way to prove that there are no others. An informal proof is given by John McClellan in his article on "The Hexahedra Problem," *Recreational Mathematics Magazine,* No. 4, August 1961, pages 34–40.

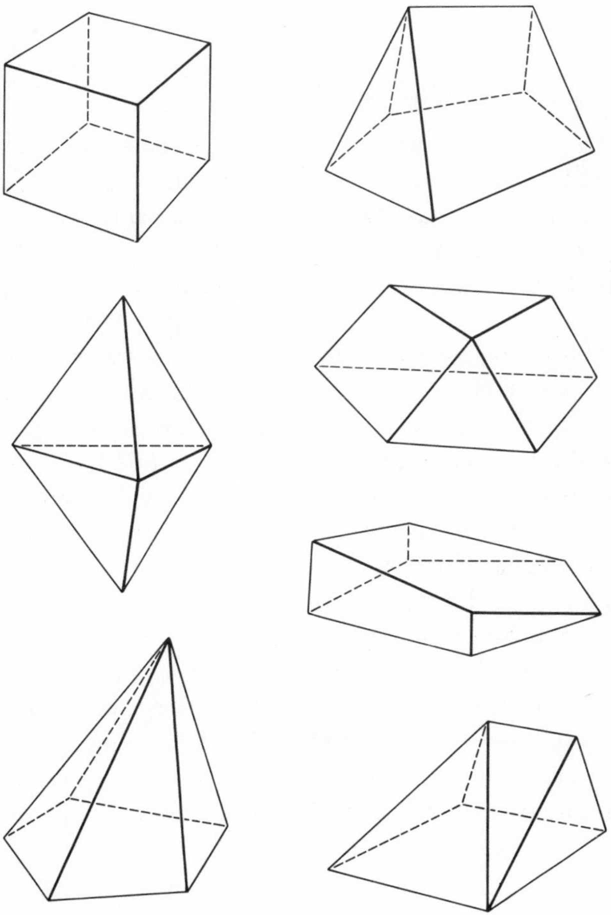

FIG. 124
The seven varieties of convex hexahedrons.

CHAPTER TWENTY

□

The Calculus of Finite Differences

THE CALCULUS of finite differences, a branch of mathematics that is not too well known but is at times highly useful, occupies a halfway house on the road from algebra to calculus. W. W. Sawyer, a mathematician at Wesleyan University, likes to introduce it to students by performing the following mathematical mind-reading trick.

Instead of asking someone to "think of a number" you ask him to "think of a formula." To make the trick easy, it should be a quadratic formula (a formula containing no powers of x greater than x^2). Suppose he thinks of $5x^2 + 3x - 7$. While your back is turned so that you cannot see his calculations, ask him to substitute 0, 1 and 2 for x, then tell you the three values that result for the entire expression. The values he gives you are -7, 1, 19. After a bit of scribbling (with practice you can do it in your head) you tell him the original formula!

The method is simple. Jot down in a row the values given to you. In a row beneath write the differences between adjacent pairs of numbers, always subtracting the number on the left from its neighbor on the right. In a third row put the difference between the numbers above it. The chart will look like this:

$$\begin{array}{ccccc} -7 & & 1 & & 19 \\ & 8 & & 18 & \\ & & 10 & & \end{array}$$

The coefficient of x^2, in the thought-of formula, is always half the bottom number of the chart. The coefficient of x is obtained by taking half the bottom number from the first number of the middle row. And the constant in the formula is simply the first number of the top row.

What you have done is something analogous to integration in calculus. If y is the value of the formula, then the formula expresses a function of y with respect to x. When x is given values in a simple arithmetic progression (0, 1, 2 . . .), then y assumes a series of values (-7, 1, 19 . . .). The calculus of finite differences is the study of such series. In this case, by applying a simple technique to three terms of a series, you were able to deduce the quadratic function that generated the three terms.

The calculus of finite differences had its origin in *Methodus Incrementorum,* a treatise published by the English mathematician Brook Taylor (who discovered the "Taylor theorem" of calculus) between 1715 and 1717. The first important work in English on the subject (after it had been developed by Leonhard Euler and others) was published in 1860 by George Boole, of symbolic-logic fame. Nineteenth-century algebra textbooks often included a smattering of the calculus, then it dropped out of favor except for its continued use by actuaries in checking annuity tables and its occasional use by scientists for finding formulas and interpolating values. Today, as a valuable tool in statistics and the social sciences, it is back in fashion once more.

For the student of recreational mathematics there are elementary procedures in the calculus of finite differences that can be enormously useful. Let us see how such a procedure can be applied to the old problem of slicing a pancake. What is the maximum number of pieces into which a pancake can be cut by n straight cuts, each of which crosses each of the others? The number is clearly a function of n. If the function is not too complex, the method of differences may help us to find it by empirical techniques.

No cut at all leaves one piece, one cut produces two pieces, two cuts yield four pieces, and so on. It is not difficult to find by trial and error that the series begins: 1, 2, 4, 7, 11 . . . [*see Fig. 125*]. Make a chart as before, forming rows, each representing the differences of adjacent terms in the row above:

Number of cuts	0		1		2		3		4
Number of pieces	1		2		4		7		11
First differences		1		2		3		4	
Second differences			1		1		1		

If the original series is generated by a linear function, the numbers in the row of first differences will be all alike. If the function is a quadratic, identical numbers appear in the row of second differences. A cubic formula (no powers higher than x^3) will have identical numbers in the row of third differences, and so on. In other words, the number of rows of differences is the order of the formula. If the chart required ten rows of differences before the numbers in a row became the same, you would know that the generating function contained powers as high as x^{10}.

Here there are only two rows, so the function must be a quadratic. Because it is a quadratic, we can obtain it quickly by the simple method used in the mind-reading trick.

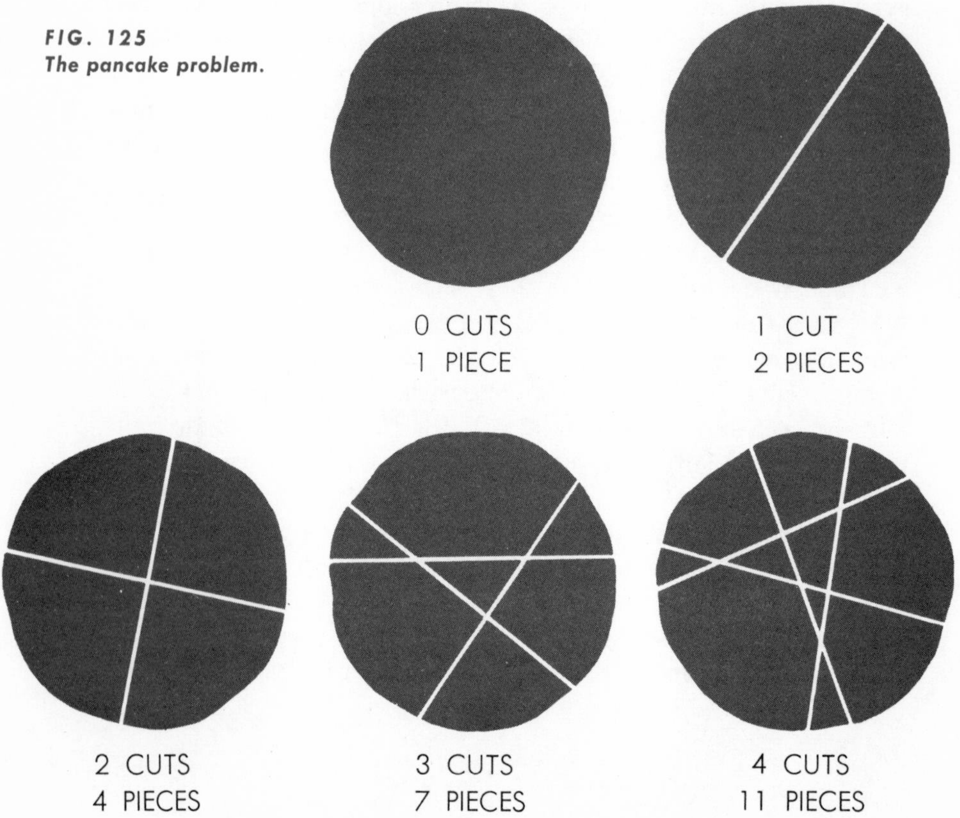

FIG. 125
The pancake problem.

The pancake-cutting problem has a double interpretation. We can view it as an abstract problem in pure geometry (an ideal circle cut by ideal straight lines) or as a problem in applied geometry (a real pancake cut by a real knife). Physics is riddled with situations of this sort that can be viewed in both ways, and that involve formulas obtainable from empirical results by the calculus of finite differences. A famous example of a quadratic formula is the formula for the maximum number of electrons that can occupy each "shell" of an atom. Going outward from the nucleus, the series runs: 0, 2, 8, 18, 32, 50. . . . The first row of differences is: 2, 6, 10, 14, 18. . . . The second row: 4, 4, 4, 4. . . . Applying the key to the mind-reading trick, we obtain the simple formula $2n^2$ for the maximum number of electrons in the nth shell.

What do we do if the function is of a higher order? We can make use of a remarkable formula discovered by Isaac Newton. It applies in all cases, regardless of the number of tiers in the chart.

Newton's formula assumes that the series begins with the value of the function when n is 0. We call this number a. The first number of the first row of differences is b, the first number of the next row is c, and so on. The formula for the nth number of the series is:

$$a + bn + \frac{cn(n-1)}{2} + \frac{dn(n-1)(n-2)}{2\cdot 3} + \frac{en(n-1)(n-2)(n-3)}{2\cdot 3\cdot 4} \ldots$$

The formula is used only up to the point at which all further additions would be zero. For example, if applied to the pancake-cutting chart, the values of 1, 1, 1 are substituted for a, b, c in the formula. (The rest of the formula is ignored because all lower rows of the chart consist of zeros; d, e, f . . . therefore have values of zero, consequently the entire portion of the formula containing these terms adds up to zero.) In this way we obtain the quadratic function $\frac{1}{2}n^2 + \frac{1}{2}n + 1$.

Does this mean that we have now found the formula for the maximum number of pieces produced by n slices of a pancake? Unfortunately the most that can be said at this point is "Probably." Why the uncertainty? Because for any finite series of numbers there is an infinity of functions that will generate them. (This is the same as saying that for any finite number of points on a graph, an infinity of curves can be drawn through those points.) Consider the series 0, 1, 2, 3. . . . What is the next term? A good guess is 4. In fact, if we apply the technique just explained, the row of first differences will be 1's, and Newton's formula will tell us that the nth term of the series is simply n. But the formula

$$n + \frac{1}{24}n(n-1)(n-2)(n-3)$$

also generates a series that begins 0, 1, 2, 3. . . . In this case the series continues, not 4, 5, 6 . . . but 5, 10, 21. . . .

There is a striking analogy here with the way laws are discovered in science. In fact, the method of differences can often be applied to physical phenomena for the purpose of guessing a natural law. Suppose, for example, that a physicist is investigating for the first time the way in which bodies fall. He observes that after one second a stone drops 16 feet, after two seconds 64 feet, and so on. He charts his observations like this:

```
0      16      64      144      256
    16      48      80      112
        32      32      32
```

Actual measurements would not, of course, be exact, but the numbers in the last row would not vary much from 32, so the physicist assumes that the next row of differences consists of zeros. Applying Newton's formula, he concludes that the total distance a stone falls in n seconds is $16n^2$. But there is nothing certain about this law. It represents no more than the simplest function that accounts for a finite series of observations; the lowest order of curve that can be drawn through a finite series of points on a graph. True, the law is confirmed to a greater degree as more observations are made, but there is never certainty that more observations will not require modification of the law.

With respect to pancake-cutting, even though a pure mathematical structure is being investigated rather than the behavior of nature, the situation is surprisingly similar. For all we now know, a fifth slice may not produce the sixteen pieces predicted by the formula. A single failure of this sort will explode the formula, whereas no finite number of successes, however large, can positively establish it. "Nature," as George Polya has put it, "may answer Yes or No, but it whispers one answer and thunders the other. Its Yes is provisional, its No is definitive." Polya is speaking of the world, not abstract mathematical structure, but it is curious that his point applies equally well to the guessing of functions by the method of differences. Mathematicians do a great deal of guessing, along lines that are often similar to methods of induction in science, and Polya has written a fascinating work, *Mathematics and Plausible Reasoning,* about how they do it.

Some trial-and-error testing, with pencil and paper, shows that five cuts of a pancake do in fact produce a maximum of sixteen pieces. This successful prediction by the formula adds to the probability that the formula is correct. But until it is rigorously *proved* (in this case it is not hard to do) it stands only as a good bet. Why the simplest formula is so often the best bet, both in mathematical and scientific guessing, is one of the lively controversial questions in contemporary philosophy of science. For one thing, no one is sure just what is meant by "simplest formula."

Here are a few problems that are closely related to pancake-cutting and that are all approachable by way of the calculus of finite differences. First you find the best guess for a formula, then you try to prove the formula by deductive methods. What is the maximum number of pieces that can be produced by n simultaneous straight cuts of a flat figure shaped like a crescent moon? How many pieces of cheesecake can be produced by n simultaneous plane cuts of a cylindrical cake? Into how many parts can the plane be divided by intersecting circles of the same size? Of different sizes? By intersecting ellipses of different sizes? Into how many regions can space be divided by intersecting spheres?

Recreational problems involving permutations and combinations often contain low-order formulas that can be correctly guessed by the method of finite differences and later (one hopes) proved. With an unlimited supply of toothpicks of n different colors, how many different triangles can be formed on a flat surface, using three toothpicks for the three sides of each triangle?

(Reflections are considered different, but not rotations.) How many different squares? How many different tetrahedrons can be produced by coloring each face a solid color and using n different colors? (Two tetrahedrons are the same if they can be turned and placed side by side so that corresponding sides match in color.) How many cubes with n colors?

Of course, if a series is generated by a function other than a polynomial involving powers of the variable, then other techniques in the method of differences are called for. For example, the exponential function 2^n produces the series 1, 2, 4, 8, 16. . . . The row of first differences is also 1, 2, 4, 8, 16 . . . , so the procedure explained earlier will get us nowhere. Sometimes a seemingly simple situation will involve a series that evades all efforts to find a general formula. An annoying example is the necklace problem posed in one of Henry Ernest Dudeney's puzzle books. A circular necklace contains n beads. Each bead is black or white. How many different necklaces can be made with n beads? Starting with no beads, the series is 0, 2, 3, 4, 6, 8, 13, 18, 30. . . . (Figure 126 shows the eighteen different varieties of necklace when $n = 7$.) I suspect that two formulas are interlocked here, one for odd n, one for even, but whether the method of differences

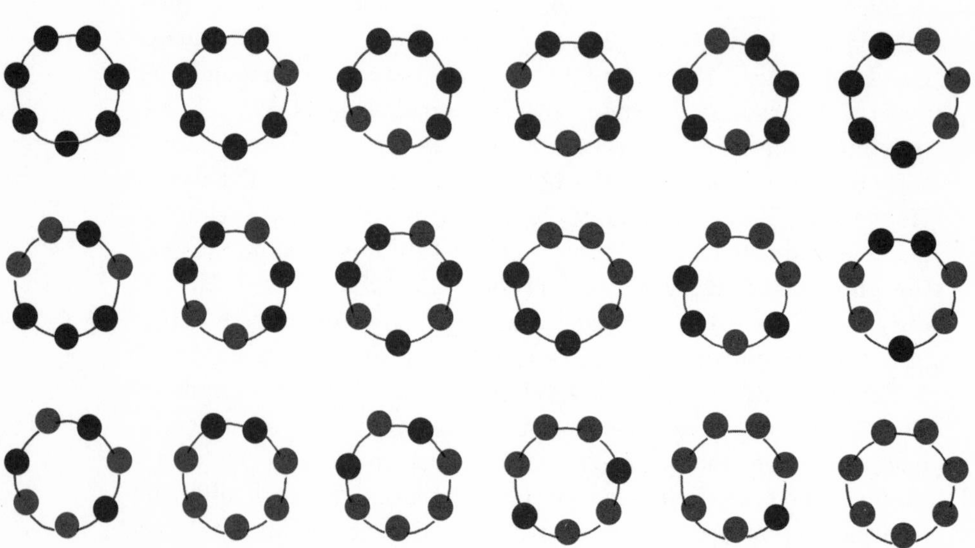

FIG. 126
Eighteen different seven-beaded necklaces can be formed with beads of two colors.

will produce the formulas, I do not know. "A general solution . . . is difficult, if not impossible," writes Dudeney. The problem is equivalent to the following one in information theory: What is the number of different binary code words of a given length, ruling out as identical all those words that have the same cyclic order of digits, taking them either right to left or left to right?

A much easier problem on which readers may enjoy testing their skill was sent to me by Charles B. Schorpp and Dennis T. O'Brien, of the Novitiate of St. Isaac Jogues in Wernersville, Pennsylvania: What is the maximum number of triangles that can be made with n straight lines? Figure 127 shows how ten triangles can be formed with five lines. How many can be made with six lines and what is the general formula? The formula can first be found by the method of differences; then, with the proper insight, it is easy to show that the formula is correct.

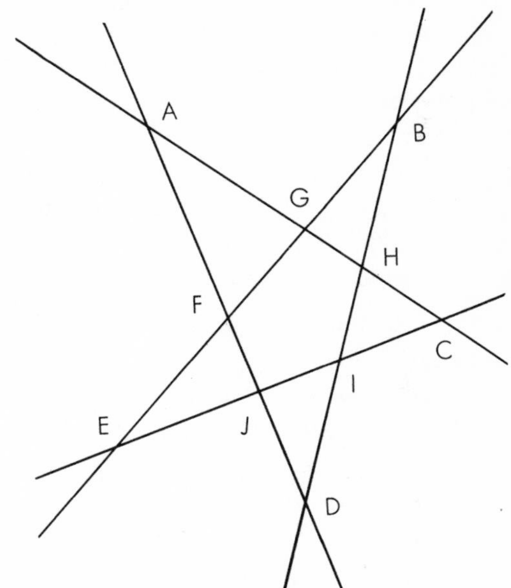

FIG. 127
Five lines make ten triangles.

ADDENDUM

IN APPLYING Newton's formula to empirically obtained data, one sometimes comes up against an anomaly for the zero case. For instance, *The Scientific American Book of Mathematical Puzzles & Diversions,* page 149, gives the formula for the maximum number of pieces that can be produced by n simultaneous plane cuts through a doughnut. The formula is a cubic

$$\frac{n^3 + 3n^2 + 8n}{6}$$

that can be obtained by applying Newton's formula to results obtained empirically, but it does not seem to apply to the zero case. When a doughnut is not cut at all, clearly there is one piece, whereas the formula says there should be no pieces. To make the formula applicable, we must define "piece" as part of a doughnut *produced by cutting*. Where there is ambiguity about the zero case, one must extrapolate backward in the chart of differences and assume for the zero case a value that produces the desired first number in the last row of differences.

To prove that the formula given for the maximum number of regions into which a pancake (circle) can be divided by n straight cuts, consider first the fact that each nth line crosses $n-1$ lines. The $n-1$ lines divide the plane into n regions. When the nth line crosses these n regions, it cuts each region into two parts, therefore every nth line adds n regions to the total. At the beginning there is one piece. The first cut adds one more piece, the second cut adds two more pieces, the third cut adds three more, and so on up to the nth cut which adds n pieces. Therefore the total number of regions is $1 + 1 + 2 + 3 + \ldots + n$. The sum of $1 + 2 + 3 + \ldots + n$ is $\frac{1}{2}n(n-1)$. To this we must add 1 to obtain the final formula.

The bead problem was given by Dudeney as problem 275 in his *Puzzles and Curious Problems*. John Riordan mentions the problem on page 162, problem 37, of his *Introduction to Combinatorial Analysis* (Wiley, 1958), indicating the solution without giving actual formulas. (He had earlier discussed the problem in "The Combinatorial Significance of a Theorem of Polya," *Journal of the Society for Industrial and Applied Mathematics,* Vol. 5, No. 4, December 1957, pages 232–234.) The problem was later treated in considerable detail, with some surprising applications to music theory and switching theory, by Edgar N. Gilbert and John Riordan, in "Symmetry Types of Periodic Sequences," *Illinois Journal of Mathematics,* Vol. 5, No. 4, December 1961, pages 657–665. The authors give the following table for the number of different types of necklaces, with beads of two colors, for necklaces of one through twenty beads:

Number of Beads	*Number of Necklaces*
1	2
2	3
3	4
4	6
5	8
6	13
7	18
8	30
9	46
10	78
11	126
12	224
13	380
14	687
15	1,224
16	2,250
17	4,112
18	7,685
19	14,310
20	27,012

The formulas for the necklace problem do not mean, by the way, that Dudeney was necessarily wrong in saying that a solution was not possible, since he may have meant only that it was not possible to find a polynomial expression for the number of necklaces as a function of n so that the number could be calculated directly from the formula without requiring a tabulation of prime factors. Because the formulas include Euler's phi function, the number of necklaces has to be calculated recursively. Dudeney's language is not precise, but it is possible that he would not have considered recursive formulas a "solution." At any rate, the calculus of finite differences is not in any way applicable to the problem, and only the recursive formulas are known.

Several dozen readers (too many for a listing of names) sent correct solutions to the problem before Golomb's formulas were printed, some of them deriving it from Riordan, others working it out entirely for themselves. Many pointed out that when the number of beads is a prime (other than 2), the formula for the number of different necklaces becomes very simple:

$$\frac{2^{n-1}-1}{n}+2^{\frac{n-1}{2}}+1$$

The following letter from John F. Gummere, headmaster of William Penn Charter School, Philadelphia, appeared in the letters department of *Scientific American* in October 1961:

Sirs:

I read with great interest your article on the calculus of finite differences. It occurs to me that one of the most interesting applications of Newton's formula is one I discovered for myself long before I had reached the calculus. This is simply applying the method of finite differences to series of powers. In experimenting with figures, I noticed that if you wrote a series of squares such as 4, 9, 16, 25, 36, 49 and subtracted them from each other as you went along, you got a series that you could similarly subtract once again and come up with a finite difference.

So then I tried cubes and fourth powers and evolved a formula to the effect that if n *is the power, you must subtract* n *times, and your constant difference will be factorial* n. *I asked my father about this (he was for many years director of the Strawbridge Memorial Observatory at Haverford College and teacher of mathematics). In good Quaker language he said: "Why, John, thee has discovered the calculus of finite differences."*

ANSWERS

HOW MANY different triangles can be formed with n straight lines? It takes at least three lines to make one triangle, four lines will make four triangles, five lines will make ten triangles. Applying the calculus of finite differences, one draws up the table in Figure 128.

The three rows of differences indicate a cubic function. Using Newton's formula, the function is found to be: $\frac{1}{6}n(n-1)(n-2)$. This will generate the series 0, 0, 0, 1, 4, 10 . . . and therefore has a good chance of being the formula for the maximum num-

NUMBER OF LINES	0	1	2	3	4	5
NUMBER OF TRIANGLES	0	0	0	1	4	10
FIRST DIFFERENCES	0	0	1	3	6	
SECOND DIFFERENCES	0	1	2	3		
THIRD DIFFERENCES	1	1	1			

FIG. 128
The answer to the triangle problem.

ber of triangles that can be made with n lines. But it is still just a guess, based on a small number of pencil and paper tests. It can be verified by the following reasoning.

The lines must be drawn so that no two are parallel and no more than two intersect at the same point. Each line is then sure to intersect every other line, and every set of three lines must form one triangle. It is not possible for the same three lines to form more than one triangle, so the number of triangles formed in this way is the maximum. The problem is equivalent, therefore, to the question: In how many different ways can n lines be taken three at a time? Elementary combinatorial theory supplies the answer: the same as the formula obtained empirically.

Solomon W. Golomb, a mathematician mentioned earlier in the chapter on polyominoes, was kind enough to send me his solution to the necklace problem. The problem was to find a formula for the number of different necklaces that can be formed with n beads, assuming that each bead can be one of two colors and not counting rotations and reflections of a necklace as being different. The formula proves to be far beyond the power of the simple method of differences.

Let the divisors of n (including 1 and n) be represented by d_1, d_2, d_3 . . . For each divisor we find what is called Euler's phi function for that divisor, symbolized $\Phi(d)$. This function is the number of integers that are prime to d; that is, which have no common divisor with d. It is assumed that 1 is such an integer, but not d. Thus Φ (8) is 4, because 8 has the following four integers that are prime to it: 1, 3, 5, 7. By convention, Φ (1) is taken to be 1. Euler's phi functions for 2, 3, 4, 5, 6, 7 are 1, 2, 2, 4, 2, 6, in the same order. Let a stand for the number of different colors each bead can be. For necklaces with an odd number of beads the formula for the number of different necklaces with n beads is the one given at the top of Figure 129. When n is even, the formula is the one at the bottom of the illustration.

$$\frac{1}{2n}\left[\phi(d_1)\cdot a^{\frac{n}{d_1}} + \phi(d_2)\cdot a^{\frac{n}{d_2}}\ldots + n\cdot a^{\frac{n+1}{2}}\right]$$

$$\frac{1}{2n}\left[\phi(d_1)\cdot a^{\frac{n}{d_1}} + \phi(d_2)\cdot a^{\frac{n}{d_2}}\ldots + \frac{n}{2}\cdot(1 + a)\cdot a^{\frac{n}{2}}\right]$$

FIG. 129
Equations for the solution of the necklace problem.

The single dots are symbols for multiplication. Golomb expressed these formulas in a more compressed, technical form, but I think the above forms will be clearer to most readers. They are more general than the formulas asked for, because they apply to beads that may have any specified number of colors.

The formulas answering the other questions in the chapter are:

1. Regions of a crescent moon produced by n straight cuts:
$$\frac{n^2 + 3n}{2} + 1$$
2. Pieces of cheesecake produced by n plane cuts:
$$\frac{n^3 + 5n}{6} + 1$$
3. Regions of the plane produced by n intersecting circles:
$$n^2 - n + 2$$
4. Regions of the plane produced by n intersecting ellipses:
$$2n^2 - 2n + 2$$
5. Regions of space produced by n intersecting spheres:
$$\frac{n\,(n^2 - 3n + 8)}{3}$$
6. Triangles formed by toothpicks of n colors:
$$\frac{n^3 + 2n}{3}$$
7. Squares formed with toothpicks of n colors:
$$\frac{n^4 + n^2 + 2n}{4}$$
8. Tetrahedrons formed with sides of n colors:
$$\frac{n^4 + 11n^2}{12}$$
9. Cubes formed with sides of n colors:
$$\frac{25n^4 - 120n^3 + 209n^2 - 108n}{6}$$

REFERENCES FOR FURTHER READING

1. The Binary System

"The Logical Abacus." W. Stanley Jevons in *The Principles of Science,* Chapter 6, pages 104–105. London: Macmillan, 1874. Reprinted as a Dover paperback, 1958.

"Two." Constance Reid in *From Zero to Infinity,* Chapter 2. Thomas Y. Crowell, revised edition, 1960.

"Some Binary Games." R. S. Scorer, P. M. Grundy and C. A. B. Smith in *Mathematical Gazette,* Vol. 28, pages 96–103, 1944.

"How to Count on Your Fingers." Frederik Pohl in *Digits and Dastards.* Ballantine, 1966.

"Card Sorting and the Binary System." John Milholland in *Mathematics Teacher,* Vol. 44, pages 312–314, 1951.

"A Punch-card Adding Machine Your Pupils Can Build." Larew M. Collister in *Mathematics Teacher,* Vol. 52, pages 471–473, October 1959.

2. Group Theory and Braids

"Theory of Braids." Emil Artin in *Annals of Mathematics,* Second Series, Vol. 48, No. 1, pages 101–126, January 1947.

"Braids and Permutations." Emil Artin in *Annals of Mathematics,* Second Series, Vol. 48, pages 643–649, 1947.

"The Theory of Braids." Emil Artin in *The American Scientist,* Vol. 38, No. 1, pages 112–119, January 1950; reprinted in *Mathematics Teacher,* Vol. 52, No. 5, pages 328–333, May 1959. A nontechnical discussion of results contained in the two previous papers.

"On a String Problem of Dirac." M. H. A. Newman in *The Journal of the London Mathematical Society,* Vol. 17, Part 3, No. 67, pages 173–177, July 1942.

On Group Theory:

The Theory of Groups. Marshall Hall, Jr. Macmillan, 1959.

Groups. Georges Papy. St. Martin's Press, 1964.

The Theory of Groups: An Introduction. Joseph J. Rotman. Allyn and Bacon, 1965.

"Group Theory for School Mathematics." Richard A. Dean in *Mathematics Teacher,* Vol. 55, No. 2, pages 98–105, February 1962.

4. The Games and Puzzles of Lewis Carroll

The Lewis Carroll Picture Book. Edited by Stuart Dodgson Collingwood. Unwin, 1899. Reprinted as a Dover paperback, 1961, under the title *Diversions and Digressions of Lewis Carroll.*

Symbolic Logic and the Game of Logic. Lewis Carroll. Dover, 1958.

Pillow Problems and a Tangled Tale. Lewis Carroll. Dover, 1958.

"Lewis Carroll and a Geometrical Puzzle." Warren Weaver in *American Mathematical Monthly,* Vol. 45, pages 234–36, April 1938.

"The Mathematical Manuscripts of Lewis Carroll." Warren Weaver in the *Proceedings of the American Philosophical Society,* Vol. 98, pages 377–381, October 15, 1954.

"Lewis Carroll: Mathematician." Warren Weaver in *Scientific American,* pages 116–128, April 1956.

"Mathematics Through a Looking Glass." Margaret F. Willerding in *Scripta Mathematica,* Vol. 25, No. 3, pages 209–219, November 1960.

The Annotated Alice. Martin Gardner. Clarkson Potter, 1960. Reprinted in paper covers by Forum Books, 1963, and Penguin, 1965.

The Annotated Snark. Martin Gardner. Simon and Schuster, 1962.

5. Paper Cutting

Paper Capers. Gerald M. Loe. Chicago: Ireland Magic Co., 1955.

A Miscellany of Puzzles. Stephen Barr. Crowell, 1965. The book contains a number of new puzzles involving paper cutting and folding.

Equivalent and Equidecomposable Figures. V. G. Boltyanskii. D. C. Heath, 1963. A paperback booklet translated from a 1956 Russian edition.

Geometric Dissections. Harry Lindgren. Van Nostrand, 1964. The definitive work on the subject.

6. Board Games

A History of Board Games other than Chess. Harold James Ruthven Murray. Oxford Press, 1952.

Board and Table Games. R. C. Bell. Oxford Press, 1960.

On Rithmomachy:

"Rithmomachia, the Great Medieval Number Game." David Eugene Smith and Clara C. Eaton in *Number Games and Number Rhymes,* pages 29–38. New York: Teachers College, Columbia University, 1914. Reprinted from the *American Mathematical Monthly,* April 1911.

"Boissiére's Pythagorean Game." John F. C. Richards in *Scripta Mathematica,* Vol. 12, No. 3, pages 177–217, September 1946.

"Ye Olde Gayme of Rithmomachy." Charles Leete in Case Institute's *Engineering and Science Review,* pages 18–20, January 1960.

On Oriental Chess:

Korean Games, with Notes on the Corresponding Games of China and Japan. Stewart Culin. University of Pennsylvania, 1895. Reprinted in 1958 by Charles E. Tuttle under the title *Games of the Orient.*

A Manual of Chinese Chess. Charles F. Wilkes. San Francisco: Yamato Press, 1952.

Japanese Chess, the Game of Shogi. E. Ohara. Bridgeway (Tuttle) Press, 1958.

On Fairy Chess:

Chess Eccentricities. Major George Hope Verney. London: Longmans, Green and Co., 1885. The best reference in English.

"Fairy Chess." Maurice Kraitchik in *Mathematical Recreations,* pages 276–279. Dover, 1953.

"Variations on Chess." V. R. Parton in *The New Scientist* (an English weekly), page 607, May 27, 1965.

Les Jeux d'Echecs Non Orthodoxes. Joseph Boyer. Published by the author, Paris, 1951.

Nouveaux Jeux d'Echecs Non Orthodoxes. Joseph Boyer. Published by the author, Paris, 1954.

Les Jeux de Dames Non Orthodoxes. Joseph Boyer. Published by the author, Paris, 1956.

On Reversi:

A Handbook of Reversi. Jacques & Son, 1888. A booklet of rules, authorized by Lewis Waterman, and sold with the game.

The Handbook of Reversi. F. H. Ayres, 1889. A booklet of rules by the rival inventor John W. Mollett, issued by a rival manufacturer to sell with their version of the game.

Reversi and Go Bang. "Berkeley" (W. H. Peel). New York: F. A. Stokes Co., 1890. A 72-page book, authorized by Waterman. The best reference on the game.

Reversi. Alice Howard Cady. New York: American Sports Publishing Co., 1896. A 44-page paperback, chiefly a simplified rewrite of the previous book.

"Reversi." "Professor Hoffmann" (Angelo Lewis) in *The Book of Table Games,* pages 611–623. London: George Routledge and Sons, 1894.

7. Packing Spheres

"In the Twinkling of an Eye." Edwin B. Matzke in the *Bulletin of the Torrey Botanical Club,* Vol. 77, No. 3, pages 222–227, May 1950.

"Close Packing and Froth." H. S. M. Coxeter in the *Illinois Journal of Mathematics,* Vol. 2, No. 48, pages 746–758, 1958. The article has a bibliography of 30 earlier references.

"The Packing of Equal Spheres." C. A. Rogers in *Proceedings of the London Mathematical Society,* Vol. 8, pages 609–620, 1958.

"Covering Space with Equal Spheres." H. S. M. Coxeter in *Mathematika,* Vol. 6, pages 147–157, 1959.

"Close Packing of Equal Spheres." H. S. M. Coxeter in *Introduction to Geometry,* pages 405–411. Wiley, 1961.

"Simple Regular Sphere Packing in Three Dimensions." Ian Smalley in *Mathematics Magazine,* pages 295–300, November 1963.

Regular Figures. L. Fejes Toth. Macmillan, 1964. See pages 288–307.

8. The Transcendental Number Pi

Famous Problems of Elementary Geometry. Felix Klein. Ginn and Co., 1897. Reprinted 1930 by Stechert, and currently available as a Dover paperback.

"The History and Transcendence of Pi." David Eugene Smith in *Monographs on Topics of Modern Mathematics,* edited by J. W. A. Young. Longmans, Green, 1911; Dover paperback, 1955.

Squaring the Circle: A History of the Problem. E. W. Hobson. Cambridge, 1913; Chelsea, 1953.

"Squaring the Circle." Heinrich Tietze in *Famous Problems of Mathematics,* Chapter 5. Graylock Press, 1965 (translated from the 1959 revised German edition).

"The Long, Long Trail of Pi." Philip J. Davis in *The Lore of Large Numbers,* Chapter 17. Random House New Mathematical Library, 1961.

"The Number Pi." H. von Baravalle in *Mathematics Teacher,* Vol. 45, pages 340–348, May 1952.

"Circumetrics." Norman T. Gridgeman in *The Scientific Monthly,* Vol. 77, No. 1, pages 31–35, July 1953.

On Computing the Decimals of Pi:

Contributions to Mathematics, comprising chiefly the rectification of the circle to 607 places of decimals. William Shanks. London, 1853.

"Statistical Treatment of the Values of First 2,000 Decimal Digits of e and π Calculated on the ENIAC." N. C. Metropolis, G. Reitwiesner and J. von Neumann in *Mathematical Tables and other Aids to Computation,* Vol. 4, pages 109–111, 1950.

"The Evolution of Extended Decimal Approximations to Pi." J. W. Wrench, Jr., in *The Mathematics Teacher,* pages 644–649, December 1960.

"Calculation of Pi to 100,000 Decimals." Daniel Shanks and John W. Wrench, Jr., in *Mathematics of Computation,* Vol. 16, No. 77, pages 76–99, January 1962. Tables giving the first 100,000 decimals of pi are included.

On Hobbes vs. Wallis:

"Hobbes' Quarrels with Dr. Wallis, the Mathematician." Isaac Disraeli in *Quarrels of Authors.* London, 1814.

Hobbes. George Croom Robertson. London: William Blackwood, 1936. See pages 167–185.

The Mathematical Works of John Wallis. Joseph F. Scott. London: Taylor and Francis, 1938.

9. Victor Eigen: Mathemagician

Mathematics, Magic and Mystery. Martin Gardner. Dover, 1956.

Mathematical Magic. William Simon. Scribner's, 1964.

"On Closed Self-intersecting Curves." Hans Rademacher and Otto Toeplitz in *The Enjoyment of Mathematics,* Chapter 10. Princeton University Press, 1957.

10. The Four-Color Map Theorem

The Four Color Problem. Philip Franklin. Scripta Mathematica Library, No. 5, 1941.

What Is Mathematics? Richard Courant and Herbert Robbins. Oxford University Press, 1941. See "The Four Color Problem," pages 246–248; "The Five Color Theorem," pages 264–267.

"The Problem of Contiguous Regions, the Thread Problem, and the Color Problem." David Hilbert and S. Cohn-Vossen in *Geometry and the Imagination,* pages 333–340. Chelsea, 1952 (translated from the German edition of 1932).

"The Four-Color Problem." Hans Rademacher and Otto Toeplitz in *The Enjoyment of Mathematics*, pages 73–82. Princeton University Press, 1957.

Introduction to Geometry. H. S. M. Coxeter. Wiley, 1961. See pages 385–395.

Intuitive Concepts in Elementary Topology. Bradford Henry Arnold. Prentice-Hall, 1962. See "The Four Color Problem," pages 43–55; "The Seven Color Theorem on a Torus," pages 85–87.

"Map Coloring." Sherman K. Stein in *Mathematics: The Man-made Universe*, pages 175–199. W. H. Freeman, 1963.

"Map Coloring." Oystein Ore in *Graphs and Their Uses*. Random House New Mathematical Library, 1963. See pages 109–116.

Induction in Geometry. L. I. Golovina and I. M. Yaglom. D. C. Heath, 1963. See pages 22–44.

Famous Problems of Mathematics. Heinrich Tietze. Graylock Press, 1965 (translated from the German edition of 1959). See "On Neighboring Domains," pages 64–89; "The Four Color Problem," pages 226–242.

"Map-Coloring Problems." H. S. M. Coxeter in *Scripta Mathematica*, Vol. 23, Nos. 1–4, pages 11–25, 1957.

"Coloring Maps." Mathematics Staff of the University of Chicago in *Mathematics Teacher*, pages 546–550, December 1957.

"The Four-Color Map Problem, 1840–1890." H. S. M. Coxeter in *Mathematics Teacher*, pages 283–289, April 1959.

"The Island of Five Colors." Martin Gardner in *Future Tense*, edited by Kendell Foster Crossen. Greenberg, 1952; reprinted in *Fantasia Mathematica*, edited by Clifton Fadiman. Simon and Schuster, 1958.

13. Polyominoes and Fault-Free Rectangles

"Polyominoes." Martin Gardner in *The Scientific American Book of Mathematical Puzzles & Diversions*, Chapter 13. Simon and Schuster, 1959.

Polyominoes. Solomon W. Golomb. Scribner's, 1965. A bibliography in the back of the book covers all important earlier references in books and magazines.

14. Euler's Spoilers

"Le problème de 36 officiers." G. Tarry in *Comptes Rendu de l'Association Française pour l'Avancement de Science Naturel*, Vol. 1, pages 122–123, 1900; Vol. 2, pages 170–203, 1901.

"On the Falsity of Euler's Conjecture about the Non-Existence of Two Orthogonal Latin Squares of Order $4t + 2$." R. C. Bose and S. S. Shrikhande in *Proceedings of the National Academy of Sciences*, Vol. 45, No. 5, pages 734–737, May 1959.

"Orthogonal Latin Squares." E. T. Parker in *Proceedings of the National Academy of Sciences*, Vol. 45, No. 6, pages 859–862, June 1959.

"Major Mathematical Conjecture Propounded 177 Years Ago Is Disproved." John A. Osmundsen in *New York Times*, page 1, April 26, 1959.

"On the Construction of Sets of Mutually Orthogonal Latin Squares and the Falsity of a Conjecture of Euler." R. C. Bose and S. S. Shrikhande in *Transactions of the American Mathematical Society*, Vol. 95, pages 191–209, 1960.

"Further Results on the Construction of Mutually Orthogonal Latin Squares and the Falsity of Euler's Conjecture." R. C. Bose, S. S. Shrikhande, and E. T. Parker in *Canadian Journal of Mathematics*, Vol. 12, pages 189–203, 1960.

"Computer Study of Orthogonal Latin Squares of Order Ten." E. T. Parker in *Computers and Automation,* pages 1–3, August 1962.

"Orthogonal Tables." Sherman K. Stein in *Mathematics: the Man-made Universe,* Chapter 12. W. H. Freeman, 1963.

"Orthogonal Latin Squares." Herbert John Ryser in *Combinatorial Mathematics,* Chapter 7. Mathematical Association of America, 1963.

On Finite Projective Planes:

"Finite Arithmetic and Geometries." W. W. Sawyer in *Prelude to Mathematics,* Chapter 13. Penguin, 1955.

"Finite Planes and Latin Squares." Truman Botts in *Mathematics Teacher,* pages 300–306, May 1961.

"Finite Planes for the High School." A. A. Albert in *Mathematics Teacher,* pages 165–169, March 1962.

"The General Projective Plane and Finite Projective Planes." Harold L. Dorwart in *The Geometry of Incidence,* Section IV. Prentice-Hall, 1966.

On the Use of Graeco-Latin Squares in Experimental Design:

Analysis and Design of Experiments. H. B. Mann. Dover, 1949.

The Design of Experiments. R. A. Fisher. Hafner, 1951.

Experimental Design and Its Statistical Basis. David John Finney. University of Chicago Press, 1955.

Planning of Experiments. D. R. Cox. John Wiley & Sons, 1958.

15. The Ellipse

"The Simplest Curves and Surfaces." David Hilbert and S. Cohn-Vossen in *Geometry and the Imagination,* pages 1–24. Chelsea, 1956.

A Book of Curves. E. H. Lockwood. Cambridge University Press, 1961.

"Something New Behind the 8-Ball." Ronald Bergman in *Recreational Mathematics Magazine,* No. 14, pages 17–19, January-February 1964. On Elliptipool.

16. The 24 Color Squares and the 30 Color Cubes

New Mathematical Pastimes. Percy Alexander MacMahon. Cambridge University Press, 1921.

Das Spiel der 30 Bunten Würfel. Ferdinand Winter. Leipzig, 1934. A 128-page paperback devoted to the 30 color cubes.

Mathematical Recreations. Maurice Kraitchik. Dover, 1953. See page 312 for a game using 30 squares that exhaust arrangements of four out of five colors, and page 313 for problems involving sixteen squares that exhaust arrangements of two colors out of eight.

"Colour-cube Problem." W. R. Rouse Ball in *Mathematical Recreations and Essays.* Revised edition. Macmillan, 1960. See pages 112–114.

"Coloured Blocks" and "Constructions from Coloured Blocks." Aniela Ehrenfeucht in *The Cube Made Interesting,* pages 46–66. Pergamon Press, 1964. The book is translated from the Polish 1960 edition.

"Stacking Colored Cubes." Paul B. Johnson in *American Mathematical Monthly,* Vol. 63, No. 6, pages 392–395, June-July 1956.

"Cubeb." L. Vosburgh Lyons in *Ibidem,* No. 12, pages 8–9, December 1957.

"Colored Polyhedra: A Permutation Problem." Clarence R. Perisho in *Mathematics Teacher,* Vol. 53, No. 4, pages 253–255, April 1960.

18. Bridg-it and Other Games

On Bridg-it:

The 2nd Scientific American Book of Mathematical Puzzles & Diversions. Martin Gardner. Simon and Schuster, 1961. See pages 84–87.

"A Solution of the Shannon Switching Game." Alfred Lehman in the *Journal of the Society of Industrial Applied Mathematics,* Vol. 12, No. 4, pages 687–725, December 1964.

On Halma:

The Book of Table Games. "Professor Hoffman" (Angelo Lewis). George Routledge and Sons, 1894. See pages 604–607.

A History of Board Games. H. J. R. Murray. Oxford University Press, 1952. See pages 51–52.

20. The Calculus of Finite Differences

The Calculus of Finite Differences. Charles Jordan. Chelsea, 1947.

Numerical Calculus. William Edmunds Milne. Princeton University Press, 1949.

The Calculus of Finite Differences. L. M. Milne-Thomson. Macmillan, 1951.

An Introduction to the Calculus of Finite Differences and Difference Equations. Kenneth S. Miller. Henry Holt, 1960.

ABOUT THE AUTHOR

MARTIN GARDNER'S first two SCIENTIFIC AMERICAN *Books of Mathematical Puzzles & Diversions* have, at this writing, brought pleasure to more than 90,000 readers and have established him (we quote Clifton Fadiman) "among the classic masters in the field." The present volume is the third in the series.

His lively marginal notes for *Alice in Wonderland* and *Through the Looking Glass* (*The Annotated Alice,* published by Clarkson Potter) have delighted Carrollians.

He is a regular contributor to *Scientific American* and is the author of numerous books on scientific and mathematical subjects, among them *Fads and Fallacies in the Name of Science,* published by Dover, *The Ambidextrous Universe* (Basic Books), and *Relativity for the Million* (Pocket Books).

Mr. Gardner was born in Tulsa, Oklahoma; was graduated from the University of Chicago, where he majored in philosophy; and started his writing career as a reporter on the Tulsa *Tribune.* The Gardners and their two children live in Hastings-on-Hudson, New York.